NON-RELATIVISTIC QUANTUM ELECTRODYNAMICS

NON-RELATIVISTIC QUANTUM ELECTRODYNAMICS

W. P. Healy

Department of Mathematics
Royal Melbourne Institute of Technology
Melbourne, Victoria
Australia

1982

ACADEMIC PRESS
A Subsidiary of Harcourt Brace Jovanovich, Publishers

LONDON NEW YORK
PARIS SAN DIEGO SAN FRANCISCO SAO PAULO
SYDNEY TOKYO TORONTO

ACADEMIC PRESS INC. (LONDON) LTD
24/28 Oval Road,
London NW1 7DX

United States Edition published by
ACADEMIC PRESS INC.
111 Fifth Avenue,
New York, New York 10003

Healy, W. P.
Non-relativistic quantum electrodynamics.
1. Quantum electrodynamics
I. Title
537.6 QC680 LCCCN 82-71002

ISBN 0-12-335720-9

Typeset and printed by The Universities Press (Belfast) Ltd.

Preface

Non-relativistic quantum electrodynamics is the theory of the interaction of photons and slowly moving charged particles, such as those found in atoms and molecules under normal laboratory conditions. In recent years there has been a resurgence of interest in this subject, due in part to the development of lasers and their use as spectroscopic tools for investigating a variety of physical and chemical phenomena. This book presents a systematic account of non-relativistic quantum electrodynamics and is aimed at theoretical physicists, applied mathematicians and physical chemists who have some knowledge of classical electrodynamics and of Dirac's formulation of ordinary quantum mechanics.

The theory is derived by the method of canonical quantization and is thus related to classical electrodynamics through Bohr's correspondence principle. This method seems the most natural and easiest to use in a non-relativistic context and, historically, is the one by which the electromagnetic field was first quantized. It is true that the description of the spin of the charged particles does not emerge automatically from the resulting formalism, but this can, if desired, be added on in an *ad hoc* fashion, as in non-relativistic quantum mechanics. The canonical formulation of classical electrodynamics in the Coulomb gauge is dealt with at some length, as are the transformations of the energy and momentum densities of the system. Particular attention is drawn to the inversion of the relations defining the electromagnetic potentials that may be achieved by taking certain line integrals over the fields. These path-dependent potentials are interesting in their own right, but do not appear to be widely known. Moreover, they are closely associated with the path-dependent polarization and magnetization fields that occur in the multipolar form of the Hamiltonian, which is related to the more standard minimal-coupling form through the unitary transformation that was first carried out by E. A. Power and S. Zienau. The interpretation of this transformation is still, even now, a subject of debate in the literature. It is hoped that the treatment given here will show (i) that the correct canonical variables and consistent equations of motion are obtained by interpreting the transformation as the quantum analogue of a classical canonical transformation and (ii) that the partitioning (for the purposes of perturbation theory) of the multipolar Hamiltonian is intrinsically path-dependent, due to the path-dependence of the polarization and magnetization fields. This can be regarded as a kind of gauge dependence. The gauge invariance of physical results, however, is demonstrated for some cases.

The summation convention is used, so that a repeated Latin suffix always implies a sum. References are listed alphabetically according to the surname of the author at the end of each chapter. They are cited in the text by author name and year of publication. A list of physical constants, some theorems in vector analysis, a note on longitudinal and transverse vector fields and two operator identities appear in Appendices A, B, C and D, respectively. The reader is referred to these where necessary in the body of the book. The problems that are given at the end of many sections are meant to illustrate the theory, and in some cases to develop it further.

The work leading to this book was mostly carried out in the Research School of Chemistry, Institute of Advanced Studies, and the Department of Applied Mathematics, Faculty of Science, at the Australian National University. Some of the last section of Chapter 7 is based on work done in collaboration with Dr. R. G. Woolley. I am very grateful to Professor D. P. Craig of the Research School of Chemistry for suggesting and encouraging the project and to Mrs B. Hawkins of the Department of Applied Mathematics for typing the manuscript.

W. P. Healy
Melbourne
March 1982

Contents

Chapter 6 **Interaction of Photons and Atoms**

Chapter 7 **Path-dependent Electrodynamics**

Appendices

Chapter 1

Introduction

1.1 Non-relativistic theory

Quantum electrodynamics is the fundamental theory of the interaction of radiation and charged particles. It is founded on the hypothesis that the behaviour of the electromagnetic field under the influence of its microscopic sources is governed by the laws of quantum mechanics as well as by the Maxwell–Lorentz equations. The theory was first established by Dirac (1927) and its modern version, due mainly to Tomonaga, Schwinger, Feynman and Dyson, complies fully with the requirements of special relativity. Indeed any theory of the pure radiation field based on the source-free Maxwell–Lorentz equations must be relativistically covariant, even though it might not be expressed in a form that makes this evident. Quantum electrodynamics has, however, a well defined non-relativistic limit in so far as the motion of the sources is concerned. The non-relativistic theory is of an approximate character but involves a much simpler formalism than its relativistic counterpart. Moreover, it is applicable to a wide range of problems in physics and chemistry, particularly in the areas of spectroscopy, laser physics and intermolecular forces.

The conditions under which non-relativistic quantum electrodynamics provides an accurate description of phenomena will first be determined. It is assumed that the charged particles move at such low speeds (in the inertial frame of a fixed observer) that their masses can be considered as constant and equal to their rest masses. More specifically, since the relativistic mass of a particle with speed v and rest mass m_0 is $m_0(1-v^2/c^2)^{-1/2}$, we require that $v/c \ll 1$, c being the speed of light *in vacuo*.† This inequality generally holds for the motion of atomic particles. Thus the root mean square speed $\bar{v}$ of the electron of a hydrogen-like ion in a state with principal quantum number n is (Pauling and Wilson 1935) $2\pi Ze^2/nh$, where Ze is the nuclear charge and h is Planck's constant. If $Z=1$ and

† For the values of physical constants, see Appendix A.

$n = 1$ (hydrogen atom in its ground state), then $\bar{v}/c$ equals the fine structure constant (approximately 1/137) and the corresponding fractional increase in mass is about 3 parts in 10^5. This ratio is larger for higher values of Z but smaller for higher values of n. The variation of mass with velocity would therefore be expected to be appreciable only for the inner-shell electrons of the heavier elements. A second assumption that will be made is that the number of each type of charged particle (electron, nucleus, etc.) is conserved. This imposes a restriction on the frequency of the radiation with which the particles interact. Associated with radiation of frequency ν are discrete quanta, or photons, of energy $h\nu$. If the frequency is high enough, the energy of a single photon may be sufficient for the creation of electron–positron pairs. To exclude this possibility (which requires an energy of order mc^2, m being the electron rest mass), we must have $\nu \ll \nu_c$, where the critical frequency ν_c is defined by $h\nu_c = mc^2$ and is about 10^{20} Hz. Hard X-rays and high-energy gamma rays are thus to be omitted from consideration.

The foregoing assumptions are consistent for elementary processes in which photons and electrons are coupled. For bound electrons this follows from the conservation of energy. A radiative transition between two states of an atom in which the electrons move with non-relativistic speeds of order v involves emission or absorption of a photon with frequency ν satisfying $\nu \ll \nu_c$. For then $h\nu \sim mv^2$ or $\nu/\nu_c \sim v^2/c^2$ which, by hypothesis, is small compared to unity. For example, to photoionize a hydrogen atom in its ground state requires only ultraviolet light, of frequency about 3×10^{15} Hz. For the scattering of photons by free electrons (Compton effect) we can invoke the law of conservation of momentum. A photon density of $1/\delta V$ per unit volume corresponds to an energy flux $h\nu c/\delta V$ and thus to a momentum density $h\nu/(c\,\delta V)$. The momentum transfer $m\,\Delta v$ to an electron by a single photon is therefore at most $h\nu/c$, so that $\Delta v/c \sim \nu/\nu_c$. This implies that an electron that was non-relativistic initially remains so after a collision with a photon of frequency much less than ν_c.

1.2 Planck's radiation law

The origins of quantum electrodynamics, as of quantum mechanics itself, may be traced to the historic investigations into the spectrum of black-body radiation which were carried out by Planck about the beginning of this century. A black body is one that absorbs all the electromagnetic energy falling upon it. The radiation emitted by such a body depends only on the absolute temperature T and may be examined by confining the system in a cavity with reflecting walls until a condition of thermal equilibrium is reached. The intensity (erg cm^{-3} Hz^{-1}) of the radiation is then given by the formula (Planck 1900a, b)

$$I_{\mathrm{P}}(\nu, T)=\frac{8\pi h\nu^3}{c^3}\frac{1}{e^{h\nu/kT}-1} \tag{1.1}$$

where k is Boltzmann's constant. This expression for the intensity conflicts with that deduced by purely classical arguments, except when $h\nu \ll kT$. The classical law, due to Rayleigh (1900) and Jeans (1905), is the limit of Equation (1.1) as h tends to zero,

$$I_{\mathrm{RJ}}(\nu, T)=\frac{8\pi\nu^2}{c^3}kT \tag{1.2}$$

and results in the so-called ultraviolet catastrophe, since the energy density, i.e. the integral of the intensity over all frequencies, does not converge. Planck's law, on the other hand, leads to a finite integrated intensity:

$$\begin{aligned}\int_0^\infty I_{\mathrm{P}}(\nu, T)\,\mathrm{d}\nu &= \frac{48\pi k^4T^4}{h^3c^3}\sum_{r=1}^{\infty}\frac{1}{r^4}\\ &= \frac{8\pi^5k^4}{15h^3c^3}T^4\end{aligned} \tag{1.3}$$

This follows from expanding the integrand in a binomial series and integrating term by term. The total intensity varies as the fourth power of the absolute temperature and thus in accordance with the Stefan–Boltzmann law. Planck's formula also reproduces Wien's displacement law, namely that the peak in the intensity distribution occurs at a frequency that is directly proportional to T. The Stefan–Boltzmann law and Wien's displacement law must hold for any distribution of the form $T^3F(u)$ (where u is the dimensionless variable $h\nu/kT$), provided the integral of $F(u)$ from 0 to ∞ exists and $F(u)$ has a maximum for some value of its argument. Thus Wien's radiation law (Wien 1896), which is represented by the equation

$$I_{\mathrm{W}}(\nu, T)=\frac{8\pi h\nu^3}{c^3}e^{-h\nu/kT} \tag{1.4}$$

and which agrees with Planck's law when $h\nu \gg kT$, also gives a total intensity proportional to T^4 and a maximum intensity at a frequency proportional to T. The constants of proportionality, however, differ from those predicted by Planck's law. In the case of Wien's law the total intensity is $48\pi k^4T^4/(h^3c^3)$, or about 92·4% of the total Planck intensity, and the maximum occurs when $u = 3$, whereas the Planck distribution has its maximum when u is approximately 2·82. Graphs of the functions $F(u)$ corresponding to the Rayleigh–Jeans, Planck and Wien radiation laws are shown in Fig. 1.

Planck's radiation law may be derived from the assumption that electromagnetic energy of frequency ν is not infinitely divisible but can be altered only by integral multiples of the quantum $h\nu$. Let us consider the discrete set of normal modes in which the cavity field, when subjected to appropriate boundary conditions at the walls, executes one-dimensional simple harmonic motion at the characteristic frequencies of the enclosure. For a sufficiently large cavity the number of modes per unit volume having frequency within $d\nu$ of ν is $8\pi\nu^2\, d\nu/c^3$ (Rayleigh 1900, Jeans 1905). The Rayleigh–Jeans law (1.2) is therefore consistent with the classical equipartition theorem, which states that in thermal equilibrium the average energy of each oscillator is kT. If the oscillators are quantized, however, then the Hamiltonian corresponding to a mode with frequency ν has a non-degenerate eigenvalue $nh\nu$ for every non-negative integer n. (The zero-point energy has been omitted.) When the oscillator

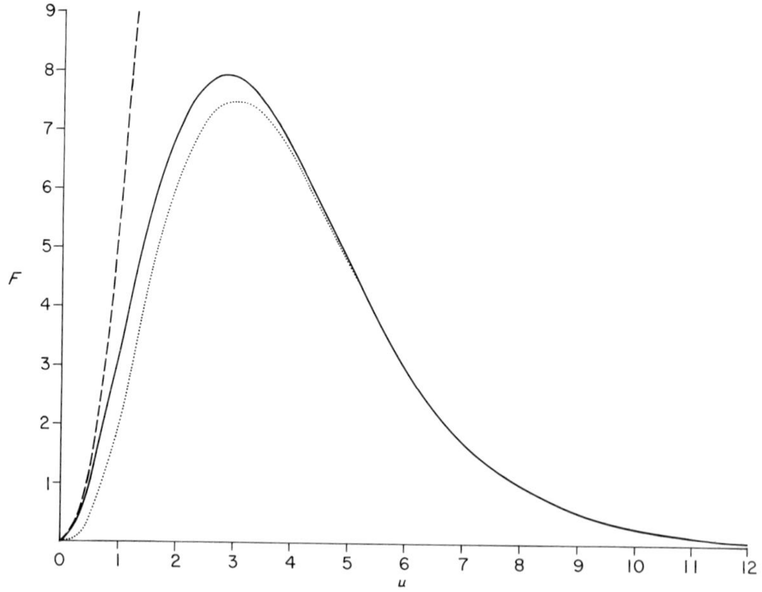

Fig. 1. Graphs of the functions $F(u)$ corresponding to the Rayleigh–Jeans (dashed curve), Planck (solid curve) and Wien (dotted curve) radiation laws. F is the intensity divided by the cube of the absolute temperature and u is the dimensionless variable $h\nu/kT$.

for such a mode is excited to state n, the field is said to contain n photons, each with energy $h\nu$. As there is no upper limit on the possible number of radiation quanta with the same frequency, polarization, etc., photons behave as indistinguishable particles obeying the Bose–Einstein statistics. The average equilibrium energies of the radiation oscillators may be calculated from the Boltzmann distribution, which determines the number of photons present. If P_n^ν denotes the probability that an oscillator of frequency ν is excited to state n, then

$$P_n^\nu(T) = N_\nu(T)e^{-nh\nu/kT} \tag{1.5}$$

where N_ν is a normalizing factor chosen so that the probabilities for a given mode sum to unity. Thus

$$N_\nu(T) = 1 - e^{-h\nu/kT} \tag{1.6}$$

Since the energy of state n is $nh\nu$ and the levels are non-degenerate, the average energy of such an oscillator is given by

$$\begin{aligned} E_\nu(T) &= \sum_{n=0}^{\infty} nh\nu P_n^\nu(T) \\ &= \frac{h\nu}{e^{h\nu/kT} - 1} \end{aligned} \tag{1.7}$$

as follows from Equations (1.5) and (1.6). Multiplication of the right-hand side of Equation (1.7) by the mode density factor $8\pi\nu^2/c^3$ then yields an energy per unit volume per unit frequency in agreement with Planck's law (1.1).

It has been shown by Lee and Stehle (1976) that the argument given above can be reversed: Planck's law and general principles of thermodynamics imply that the free field must be quantized and that photons are bosons. Similarly Wien's radiation law (1.4), which is approximately valid for high frequencies or low temperatures, admits of a particle interpretation. In this case, however, the energy levels of the radiation oscillators are degenerate and the particles no longer indistinguishable.

1.3 Einstein's *A* and *B* coefficients

The interaction of photons with atoms or molecules must, as has already been mentioned, comply with the law of conservation of energy. Suppose we have a system of identical and independent atoms (as in a rarefied gas) with energy levels $E_0, E_1, E_2, \ldots$ arranged in order of increasing magnitude. (For simplicity, we consider the discrete spectrum only.) An atom can make a transition from a state with energy E_r to one with energy E_s and simultaneously emit ($r > s$) or absorb ($r < s$) a photon of frequency ν only if

$$h\nu = |E_r - E_s| \tag{1.8}$$

This is Bohr's frequency condition (Bohr 1913). The process of emission may be spontaneous, i.e. may occur without any photons present initially, or be induced, i.e. may be due to irradiation of the atom. In the latter case the emitted photon has all the characteristics (frequency, polarization and propagation direction) of the incident light, whereas in spontaneous emission there is a probability for each direction and polarization. The rates at which emission and absorption take place were studied by Einstein (1917) on the basis of a simple model. It is assumed that during a time Δt the number of atoms spontaneously emitting a photon (of any propagation direction and polarization) while making a transition from a state with energy E_r to one with lower energy E_s is proportional to Δt and to the number of atoms with energy E_r. The constant of proportionality A_s^r then represents the emission probability per atom per unit time. The time interval Δt need not be infinitesimally short, but must be small compared to the natural lifetime of the excited state, so that the emission probability during this interval is also small. For induced emission and absorption the transition probabilities are assumed to be proportional to the intensity I of the radiation at the transition frequency ν_{rs}, as well as to the populations and the time. The corresponding coefficients B_s^r (induced emission) and B_r^s (absorption) will be independent of direction and polarization if the radiation is isotropic and unpolarized or if the atoms (or molecules) are randomly oriented and have no intrinsic chirality (i.e. are not optically active). The reciprocal time interval $1/\Delta t$ for the downward transition must now be large compared to $B_s^r I$ and that for the upward transition large compared to $B_r^s I$. The substates of any degenerate level are assumed to be equally populated (as is the case for thermal equilibrium), since otherwise the A and B coefficients would depend on the relative populations. Moreover, these coefficients include contributions from all allowed transitions between levels E_r and E_s, i.e. an averaging over initial substates and a summation over final substates have been carried out.

Certain relations between the Einstein A and B coefficients may be deduced from the conditions for thermal equilibrium between the system of atoms and the radiation. The intensity of the radiation is in this case given by Planck's law and the populations of the atomic states are determined by the Boltzmann distribution. If N_r and N_s denote the number of atoms with energy E_r and E_s, respectively, then

$$\frac{N_r(T)}{N_s(T)} = \frac{g_r}{g_s} e^{-(E_r - E_s)/kT} \tag{1.9}$$

where g_r and g_s are the statistical weights (degrees of degeneracy) of levels r and s. (It follows from this that at high temperatures the level populations are proportional to their statistical weights and that at absolute zero all atoms are in the ground state.) Equilibrium is maintained only if the number of $r \to s$ and $s \to r$ transitions per unit time are equal, since the level populations as well as the radiation energy density at the transition frequency ν_{rs} must be constant. This implies that

$$A_s^r N_r = (B_r^s N_s - B_s^r N_r) I_P(\nu_{rs}, T) \tag{1.10}$$

The left-hand side here represents the number of spontaneous $r \to s$ transitions per unit time and the right-hand side the number of absorptive $s \to r$ transitions minus the number of induced $r \to s$ transitions per unit time. Inserting for the intensity the expression given by Equation (1.1) and using the relation (1.9) for the ratio of the populations, we find that in the limit as T tends to infinity,

$$g_r B_s^r = g_s B_r^s \tag{1.11}$$

This relation must be universally valid, however, since the B coefficients do not depend on the temperature and are unchanged even if the equilibrium is destroyed. From Equations (1.10) and (1.11) we obtain a relation between the A and B coefficients, namely

$$A_s^r = \frac{8\pi h \nu_{rs}^3}{c^3} B_s^r \tag{1.12}$$

It may readily be verified that if the radiation remains isotropic and unpolarized and the sublevels equally populated, then the relations (1.11) and (1.12) are sufficient as well as necessary conditions for equilibrium.

The values of Einstein's A and B coefficients can be calculated only from a dynamical theory for the interaction of radiation and matter. The calculation was first carried out in the context of quantum electrodynamics by Dirac (1927) and will be presented in Chapter 6. It is possible to evaluate the B coefficients alone using the semiclassical theory of radiation (see, e.g., Schiff 1968), in which the atoms are supposed to be acted on by prescribed external classical fields. This theory is unable to account for the effect of the charges on the field and predicts that in the absence of external perturbations even an atom in an excited state is stable against radiative decay. In recent years, however, a dynamical theory that still treats the electromagnetic field classically has been developed (Jaynes and Cummings 1963, Crisp and Jaynes 1969, Stroud and Jaynes 1970). This so-called neoclassical theory does predict spontaneous emission in certain instances, but always at a rate lower than that predicted by quantum electrodynamics and thus incompatible with the conditions for thermal equilibrium (in patricular the relation (1.12)) between an ensemble of atoms and radiation (see Nash and Gordon 1975, Gordon 1975). The reason for the discrepancy is the neglect of field fluctuations in the neoclassical theory, which deals with the radiation generated by only the quantum expectation values of the atomic charge and current densities. In quantum electrodynamics, on the other hand, the effects of fluctuations are included. For example, when the quantized cavity oscillators considered earlier are all in their ground states, there is still a zero-point motion that gives rise to fluctuations in the vacuum field. These fluctuations are a direct consequence of the quantization of the field. The zero-point motion can be treated in a purely classical way by superimposing on the electromagnetic field a temperature-independent fluctuation in the form of plane waves that satisfy the free-field Maxwell equations but have random phases. In this "stochastic electrodynamics" the origin of the fluctuating field is unspecified; Planck's constant is introduced as a scaling factor to give the same energy spectrum as that of the zero-point radiation in quantum electrodynamics. The present status of stochastic electrodynamics has been discussed by Boyer (1980). The theory does not appear to be equivalent to quantum electrodynamics nor, as yet, capable of dealing with the same range of phenomena.

1.4 Uncertainty relations

Quantum electrodynamics must account not only for the corpuscular properties of radiation that manifest themselves in, e.g., the photoelectric and Compton effects, but also for the wave properties that appear in interference and diffraction experiments. These latter are adequately described by the Maxwell–Lorentz equations, which are retained as operator equations in the quantum theory of radiation. Thus if the electromagnetic field has a node due to interference at a point *P*, there will (at least in dipole approximation) be no absorption of photons by an atom located at *P*. The form of the wave appropriate to a particular phenomenon may, however, depend on the observations that are made on the system (Heisenberg 1930). Consider the recoil of magnitude $h\nu/c$ which, according to the law of conservation of momentum, must be imparted to an atom, initially at rest, through emission or absorption of a photon of frequency ν. The velocity accompanying this recoil, although small, was found by Einstein (1917) to be important for maintaining thermodynamic equilibrium between radiation and atoms, the Maxwell velocity distribution for the atoms being compatible with Planck's law for the radiation. (In the previous section the energy exchange but not the momentum transfer necessary for equilibrium was examined.) If the recoil is actually observed to take place on the emission of a photon, then this latter will have a momentum of equal magnitude and opposite direction to that of the atom. If, on the other hand, an accurate determination of the atom's position is carried out, then all knowledge of its linear momentum is lost and the question of recoil does not arise. However, the photon is now associated, not with a plane wave of a definite propagation direction, but with a spherical wave emanating from the observed position of the atom.

The interdependence of the quantum theories of radiation and matter is further illustrated by the uncertainty relations which the canonically conjugate dynamical variables that appear in the formalism must satisfy. It is well known (Bohr 1928, Heisenberg 1930) that the reciprocal relation

$$\Delta p\, \Delta q \sim h \tag{1.13}$$

between the uncertainty in a particle's position coordinate q and that in its corresponding momentum component p may be derived from an idealized experiment in which the position is measured by means of light scattered from the particle into a microscope. (This relation can also be obtained in other ways, e.g. by considering the diffraction of De Broglie waves by a slit.) If ε denotes the half-angle of the cone with vertex at the particle's position and base formed by the object lens of the microscope (see Fig. 2), then, according to the laws of classical optics, the accuracy of the position measurement depends on ε and the wavelength λ of the incident light as follows:

$$\Delta q \sim \frac{\lambda}{\sin \varepsilon} \tag{1.14}$$

According to the corpuscular theory, however, the incident light consists of discrete quanta, each with momentum h/λ. If only a single photon is involved in the measuring process, the uncertainty in the particle's momentum after the collision must be given by

$$\Delta p \sim \frac{h}{\lambda} \sin \varepsilon \tag{1.15}$$

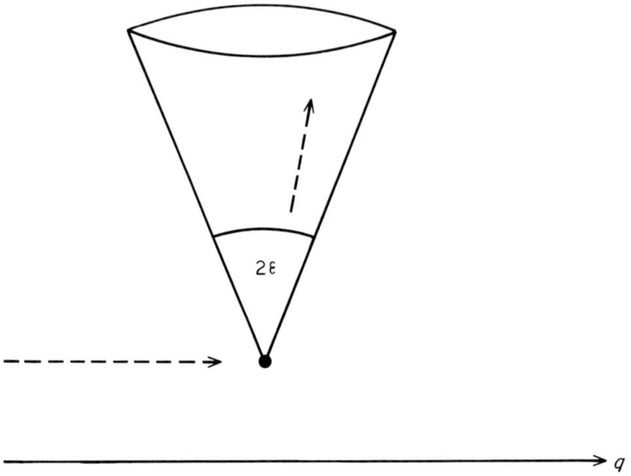

Fig. 2. Measurement of a particle's position coordinate q by means of a microscope. The dotted lines represent the incident and scattered photons and 2ε is the angle subtended at the particle's position by the lens.

since it is unknown at what angle the scattered photon enters the lens and since this angle cannot be determined without interfering destructively with the position measurement. The uncertainty Δq will be reduced if light of a shorter wavelength or a microscope with a larger aperture is used. This, however, in accordance with Bohr's complementarity principle, increases the uncertainty Δp in such a way that the product $\Delta p\,\Delta q$ always satisfies the relation (1.13) Similar conclusions are obtained if the particle's momentum is measured by observing Doppler-shifted radiation emitted from it. In the above derivation the use of the dual (wave-like and particle-like) nature of light was crucial. A classical light beam would afford a means of circumventing the uncertainty relation for particles (Heitler 1954), since the wavelength could be held fixed but the intensity decreased to such an extent that the momentum transfer to the particle becomes as small as is desired.

The dynamical variables of the radiation field also are subject to uncertainty relations (involving Planck's constant h) which may be derived from a consideration of the measurement of the field strengths by means of charged particles. The measurement problem has been discussed in detail by Bohr and Rosenfeld (1933, 1950). The quantum mechanical uncertainties associated with the charged particles used to measure the field place a limit on the accuracy with which the average values of two field strengths over certain space-time regions can be simultaneously known. (Since a field component at a definite point in space and a definite instant of time appears an abstraction from physical reality, only such average values need be considered.) The average value $\bar{E}$ of a component of the electric field over a volume V and a time interval T, for instance, can be found by measuring the change produced by the field in the momentum of a charged test body occupying the volume during this time. Although the position and momentum of the test body are uncertain by amounts Δq and Δp that satisfy the relation (1.13), it may be shown (see Heitler 1954) that this does not impair the accuracy of the field measurement, provided the mass and charge of the test body can be chosen at will. Choosing a sufficiently massive body ensures that its acceleration due to the field is negligible and thus that the body does not appreciably move from the volume V during the time T. If also the charge Q of the test body is large enough, the uncertainty in the average field, namely

$$\Delta\bar{E} \sim \frac{\Delta p}{QT} \sim \frac{h}{Q\,\Delta q T} \tag{1.16}$$

becomes very small even when Δq is small. Let us consider, however, the measurements of the average fields $\bar{E}$ and $\bar{E}'$ over two space regions V and V' during time intervals T and T', respectively. If the separation distance R between V and V' (see Fig. 3) is such that most of the light signals emitted from V during the time T will reach V' during the time T', then the measurement of $\bar{E}$ will influence that of $\bar{E}'$ in a way that is to some extent unknown. For the field produced by the test body used to measure $\bar{E}$ is superimposed on $\bar{E}'$ and cannot be fully subtracted out, as its value is uncertain (due to the uncertainty Δq in the position of the test

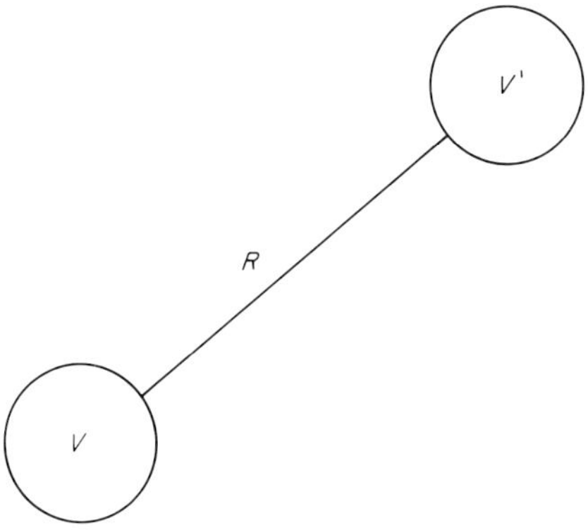

Fig. 3. Illustrating the volumes V and V' in which the average fields are measured. R is the distance separating the two regions.

body) by an amount $\Delta\bar{E}'$ equivalent to the field of a dipole of moment $Q\,\Delta q$ located in V. This field, of order $Q\,\Delta q/R^3$, can be made small by making $Q\,\Delta q$ small, but then, according to Equation (1.16), $\Delta\bar{E}$ becomes large. (This again illustrates Bohr's complementarity principle.) The order of magnitude of $\Delta\bar{E}\,\Delta\bar{E}'$ is given by

$$\Delta\bar{E}\,\Delta\bar{E}' \sim \frac{h}{R^3 T} \tag{1.17}$$

and is independent of both Q and Δq. Thus only for well-separated regions or over long intervals of time can both fields be measured with unlimited accuracy. The precise values of $\Delta\bar{E}\,\Delta\bar{E}'$ and other uncertainty products depend also on the geometry of the space-time regions VT and $V'T'$ and their relative orientation, and can be obtained either from a more detailed treatment of the measuring process or from the commutation relations satisfied by the components of the quantized field.

REFERENCES

Bohr, N. (1913). "On the constitution of atoms and molecules". *Phil. Mag.* **26,** 1.

Bohr, N. (1928). "The quantum postulate and the recent development of atomic theory". *Nature* **121,** 580.

Bohr, N. and Rosenfeld, L. (1933). "Zur Frage der Messbarkeit der Elektromagnetischen Feldgrössen". *Det. Kgl. Danske Videnskabernes Selskab. Mathematisk-fysiske Meddelelser* **12,** No. 8.

Bohr, N. and Rosenfeld, L. (1950). "Field and charge measurements in quantum electrodynamics". *Phys. Rev.* **78,** 794.

Boyer, T. H. (1980). *In* "Foundations of Radiation Theory and Quantum Electrodynamics" (ed. A. O. Barut). Plenum Press, New York and London.

Crisp, M. D. and Jaynes, E. T. (1969). "Radiative effects in semi-classical theory". *Phys. Rev.* **179,** 1253.

Dirac, P. A. M. (1927). "The quantum theory of the emission and absorption of radiation". *Proc. R. Soc. Lond. A* **114,** 243.

Einstein, A. (1917). "Zur Quantentheorie der Strahlung". *Physikalische Zeitschrift* **18,** 121.

Gordon, J. P. (1975). "Neoclassical physics and blackbody radiation". *Phys. Rev. A* **12,** 2487.

Heisenberg, W. (1930). "The Physical Principles of the Quantum Theory". University of Chicago Press.

Heitler, W. (1954). "The Quantum Theory of Radiation". Clarendon Press, Oxford.

Jaynes, E. T. and Cummings, F. W. (1963). "Comparison of quantum and semiclassical radiation theories with application to the beam maser". *Proc. I.E.E.E.* **51,** 89.

Jeans, J. H. (1905). "On the partition of energy between matter and aether". *Phil. Mag.* **10,** 91.

Lee, H. W. and Stehle, P. (1976). "Quantization of the free field". *Phys. Rev. Lett.* **36,** 277.
Nash, F. R. and Gordon, J. P. (1975). "Implications of radiative equilibrium in neoclassical theory". *Phys. Rev. A* **12,** 2472.
Pauling, L. and Wilson, E. B. (1935). "Introduction to Quantum Mechanics". McGraw-Hill, New York.
Planck, M. (1900a). "Ueber eine Verbesserung der Wien'schen Spectralgleichung". *Verhandl. der Deutschen Physikal. Gesellsch* **2,** 202.
Planck, M. (1900b). "Zur Theorie des Gesetzes der Energieverteilung im Normalspectrum". *Verhandl. der Deutschen Physikal. Gesellsch.* **2,** 237.
Rayleigh, Lord (1900). "Remarks upon the law of complete radiation". *Phil. Mag.* **49,** 539.
Schiff, L. I. (1968). "Quantum Mechanics". McGraw-Hill, New York.
Stroud, C. R. and Jaynes, E. T. (1970). "Long-term solutions in semiclassical radiation theory". *Phys. Rev. A* **1,** 106.
Wien, W. (1896). "Ueber die Energievertheilung im Emissionsspectrum eines schwarzen Körpers". *Annalen der Physik und Chemie (Leipzig)* **58,** 662.

Chapter 2

The Classical Equations of Motion

2.1 Maxwell–Lorentz theory

Since the non-relativistic quantum theory of radiation has its roots in classical electrodynamics, we shall begin our exposition with an account of this latter theory. The physical system to be considered consists of a fixed number of charged particles in interaction with the electromagnetic field. The particles will be supposed to be either electrons or nuclei, and they may occur individually in asymptotically free states or be bound together in stable aggregates such as atoms or molecules. The classical theory that describes this system is based on the Maxwell–Lorentz equations. The original electromagnetic field theory of Maxwell dealt only with the macroscopic manifestations of matter. Lorentz later postulated that the fundamental interactions take place between microscopic particles (which he called electrons) and locally fluctuating fields, and he then *derived*, by means of a statistical averaging procedure, the Maxwell theory for macroscopically observable fields and quasi-continuous matter. We shall be concerned for the most part with the microscopic description only, so that the subject to be developed here can be regarded as the quantum mechanical extension of the electron theory of Lorentz.

The partial differential equations of Maxwell and Lorentz (Maxwell 1891, Lorentz 1915) are local equations that relate the electric field $\boldsymbol{e}$ and magnetic induction field $\boldsymbol{b}$ to the charge and current densities ρ and $\boldsymbol{j}$. These equations are

$$\nabla \cdot \boldsymbol{e} = 4\pi\rho \tag{2.1}$$

$$\nabla \cdot \boldsymbol{b} = 0 \tag{2.2}$$

$$\nabla \times \boldsymbol{e} = -\frac{1}{c}\frac{\partial \boldsymbol{b}}{\partial t} \tag{2.3}$$

$$\nabla \times \boldsymbol{b} = \frac{4\pi}{c}\boldsymbol{j} + \frac{1}{c}\frac{\partial \boldsymbol{e}}{\partial t} \tag{2.4}$$

Here all fields are evaluated at the point $\boldsymbol{x}$ and the time t. The microscopic fields $\boldsymbol{e}$ and $\boldsymbol{b}$ are denoted by lower-case letters to distinguish them from their macroscopic counterparts $\boldsymbol{E}$ and $\boldsymbol{B}$. The microscopic charge and current densities ρ and $\boldsymbol{j}$ arise from a collection of material particles which will be regarded simply as geometrical points endowed with mass and charge. If e_α is the charge of particle α and $\boldsymbol{q}_\alpha(t)$ its position at time t, then the explicit expressions for ρ and $\boldsymbol{j}$ are

$$\rho(\boldsymbol{x}, t) = \sum_\alpha e_\alpha \delta(\boldsymbol{x} - \boldsymbol{q}_\alpha) \tag{2.5}$$

and

$$\boldsymbol{j}(\boldsymbol{x}, t) = \sum_\alpha e_\alpha \dot{\boldsymbol{q}}_\alpha \delta(\boldsymbol{x} - \boldsymbol{q}_\alpha) \tag{2.6}$$

where δ is Dirac's delta function and the dot denotes differentiation with respect to time. The charge and current densities thus vanish everywhere except at the particle positions, where they are infinite. Equations (2.1) and (2.4) are readily seen to be consistent with the equation of continuity or charge conservation, namely

$$\nabla \cdot \boldsymbol{j} + \frac{\partial \rho}{\partial t} = 0 \tag{2.7}$$

This equation may be deduced directly from the expressions (2.5) and (2.6) for ρ and $\boldsymbol{j}$. Implicit in the Maxwell–Lorentz equations are the microscopic analogues of the following physical laws, which were first established in the macroscopic régime:

(i) Gauss' flux law for the electric field, equivalent to Coulomb's law in the static or near zone limit.

(ii) Gauss' flux law for the magnetic induction field taken in conjunction with the non-existence of magnetic monopoles in nature.

(iii) Faraday's law of electromagnetic induction, the negative sign in Equation (2.3) being a consequence of Lenz's law.

(iv) Ørsted's law for the magnetic effect of an electric current, this latter being supplemented by Maxwell's vacuum displacement current to give a divergence-free total current and to account for the phenomenon of electromagnetic waves.

These laws describe the effect of the charged particles on the electromagnetic field. The effect of the field on the particles is determined by Newton's second law of motion with the electromagnetic force of Lorentz,

$$m_\alpha \ddot{\boldsymbol{q}}_\alpha = e_\alpha \left(\boldsymbol{e} + \frac{1}{c} \dot{\boldsymbol{q}}_\alpha \times \boldsymbol{b} \right) \tag{2.8}$$

In this equation m_α is the mass of particle α and the fields $\boldsymbol{e}$ and $\boldsymbol{b}$ are evaluated at the instantaneous position of this particle. For a system of static point charges the Lorentz force reduces to the Coulomb force, and for charges in steady motion in a static magnetic field the Lorentz force gives the mechanical effect summarized by Ampere's law for the forces between current-carrying conductors.

The coupled equations (2.1)–(2.4) and (2.8) are the equations of motion for the complete dynamical system consisting of the electromagnetic field and charged matter. Solutions satisfying specified boundary or initial conditions can be found for either (i) the electromagnetic field due to a prescribed charge–current distribution or (ii) the motion of a charged particle in a prescribed electromagnetic field. It is not known, however, whether there are any exact solutions to the set of *coupled* equations for the interacting system, or even whether these equations are mutually consistent. In later chapters we shall obtain, by means of perturbation theory, some approximate solutions to the quantum mechanical form of the coupled equations of motion.

2.2 *C, P* and *T* symmetries

The Maxwell–Lorentz equations are in agreement with the principles of special relativity, since they are covariant (i.e. retain their form) under the extended group of inhomogeneous Lorentz transformations, provided only that the combined charge and current densities transform as a four-vector and the combined electric and magnetic induction fields as an antisymmetric four-tensor (see, e.g., Bergmann 1942). It is to be noted that these equations do not involve the masses of the material particles. The Newtonian law of motion given by Equation (2.8), on the other hand, is a non-relativistic approximation, as this law is valid in this form only if the particle masses can be regarded as constant, i.e., only if the particles are observed from an inertial reference frame in which their speed is much less than that of light. The complete scheme of coupled equations is therefore not covariant under either the extended Poincaré group of special relativity or the extended Galilei group of classical mechanics. If, however, we start from any inertial frame in which the non-relativistic approximation for the material particles is valid, then the coupled equations are covariant under *the subgroup of Lorentz transformations that are also Galilei transformations*. This subgroup is generated by the following transformations:

(i) space rotations;
(ii) space displacements and space inversion;
(iii) time displacements and time reversal.

The space rotations, space displacements and time displacements together generate a further subgroup consisting of proper transformations which are continuously connected to the identity. The symmetries associated with them lead to the laws of conservation of angular momentum, linear momentum and energy, and will be discussed in Chapter 3. (The energy and momentum balances for a finite region of space, however, are discussed below in Section 2.6.) In this section we deal with the discrete and improper transformations of space inversion P and time reversal T together with the charge-conjugation transformation C.

To formulate the effect of the C, P and T transformations, we adopt the passive (or alias) point of view rather than the active (or alibi) point of view. This means that we consider a single physical system described by different observers related by the above transformations, rather than a single observer describing different physical systems related by the transformations. Thus C connects two observers who use the same space-time coordinate frame and ascribe the same magnitude to the charges of the particles but have opposing conventions for their signs.† Similarly P connects two observers who agree on the values of the charges and the time of any physical event but whose spatial axis systems, while having the same origin and unit of length, are inverted with respect to each other. Finally T connects two observers who use the same charge convention and the same spatial reference frame but whose clocks, while having a common rate and origin, have opposing senses, so that those of one run backward if those of the other run forward. The C, P and T transformations defined in this way generate an Abelian group G of order eight which may be represented by the 3×3 diagonal matrices with diagonal elements ± 1. In these matrices the first, second and third diagonal elements are -1 or $+1$ according as (i) the signs of the charges are or are not changed, (ii) the spatial axis system is or is not inverted and (iii) the time is or is not reversed, respectively. If the representation matrix corresponding to a particular group element is denoted by the same symbol as the element itself and if I is the identity element, then

$$
\begin{aligned}
&I=d(1,1,1), \qquad C=d(-1,1,1), \qquad P=d(1,-1,1), \qquad T=d(1,1,-1),\\
&\bar{I}=d(-1,-1,-1), \quad \bar{C}=d(1,-1,-1), \quad \bar{P}=d(-1,1,-1),\\
&\qquad\qquad\qquad\qquad\qquad\qquad \bar{T}=d(-1,-1,1) \qquad (2.9)
\end{aligned}
$$

where $\bar{I}=CPT$, $\bar{C}=\bar{I}C$, $\bar{P}=\bar{I}P$, $\bar{T}=\bar{I}T$ and $d(a,b,c)$ denotes a diagonal matrix with diagonal elements a, b and c. This representation of the group is unitary and faithful, but it is not irreducible since it is already in

† More generally, the effect of C is to replace particle by antiparticle properties. In our case, however, since all particles are characterized by their mass and charge, C merely reverses the sign of the charge of every particle.

TABLE 1
Multiplication table for the group G generated by C, P and T

	I	$\bar{\mathrm{I}}$	C	$\bar{\mathrm{C}}$	P	$\bar{\mathrm{P}}$	T	$\bar{\mathrm{T}}$
I	I	$\bar{\mathrm{I}}$	C	$\bar{\mathrm{C}}$	P	$\bar{\mathrm{P}}$	T	$\bar{\mathrm{T}}$
$\bar{\mathrm{I}}$	$\bar{\mathrm{I}}$	I	$\bar{\mathrm{C}}$	C	$\bar{\mathrm{P}}$	P	$\bar{\mathrm{T}}$	T
C	C	$\bar{\mathrm{C}}$	I	$\bar{\mathrm{I}}$	$\bar{\mathrm{T}}$	T	$\bar{\mathrm{P}}$	P
$\bar{\mathrm{C}}$	$\bar{\mathrm{C}}$	C	$\bar{\mathrm{I}}$	I	T	$\bar{\mathrm{T}}$	P	$\bar{\mathrm{P}}$
P	P	$\bar{\mathrm{P}}$	$\bar{\mathrm{T}}$	T	I	$\bar{\mathrm{I}}$	$\bar{\mathrm{C}}$	C
$\bar{\mathrm{P}}$	$\bar{\mathrm{P}}$	P	T	$\bar{\mathrm{T}}$	$\bar{\mathrm{I}}$	I	C	$\bar{\mathrm{C}}$
T	T	$\bar{\mathrm{T}}$	$\bar{\mathrm{P}}$	P	$\bar{\mathrm{C}}$	C	I	$\bar{\mathrm{I}}$
$\bar{\mathrm{T}}$	$\bar{\mathrm{T}}$	T	P	$\bar{\mathrm{P}}$	C	$\bar{\mathrm{C}}$	$\bar{\mathrm{I}}$	I

reduced form. (Indeed, since the group is Abelian, the only irreducible representations are of dimension unity (Wigner 1959).) The group multiplication table may easily be worked out, from the representation (2.9) or otherwise, and is shown in Table 1. The effects of the transformations on the charge and current densities and the electric and magnetic induction fields follow from Equations (2.5) and (2.6) and the Maxwell–Lorentz equations, and are listed in Table 2. The transformation properties of w and $\boldsymbol{f}$, where w is the power density $\boldsymbol{j}\cdot\boldsymbol{e}$ and $\boldsymbol{f}$ the Lorentz force density $\rho\boldsymbol{e}+c^{-1}\boldsymbol{j}\times\boldsymbol{b}$, are also shown. It may be noted from the effect of the parity transformation P that ρ and w are true scalars and $\boldsymbol{j}$, $\boldsymbol{e}$ and $\boldsymbol{f}$ true vectors, but that $\boldsymbol{b}$ is a pseudovector. Those elements of G that leave the components of $\boldsymbol{e}$ and $\boldsymbol{b}$ invariant, namely I and $\bar{I}$, form a normal subgroup, $\mathscr{I}$ say, of index 4 in G. The cosets of this subgroup are given by

TABLE 2
Field transformation properties under the group G

	ρ	j_i	e_i	b_i	w	f_i
I	ρ	j_i	e_i	b_i	w	f_i
$\bar{\mathrm{I}}$	$-\rho$	$-j_i$	e_i	b_i	$-w$	$-f_i$
C	$-\rho$	$-j_i$	$-e_i$	$-b_i$	w	f_i
$\bar{\mathrm{C}}$	ρ	j_i	$-e_i$	$-b_i$	$-w$	$-f_i$
P	ρ	$-j_i$	$-e_i$	b_i	w	$-f_i$
$\bar{\mathrm{P}}$	$-\rho$	j_i	$-e_i$	b_i	$-w$	f_i
T	ρ	$-j_i$	e_i	$-b_i$	$-w$	f_i
$\bar{\mathrm{T}}$	$-\rho$	j_i	e_i	$-b_i$	w	$-f_i$

The rows give the values ascribed to the fields at a given physical space-time point by observers related through the transformations shown on the left to a standard observer, for whom the fields at the same point have the values specified in the top row.

TABLE 3
Multiplication table for the Klein four-group

	$\mathcal{I}$	$\mathcal{C}$	$\mathcal{P}$	$\mathcal{T}$
$\mathcal{I}$	$\mathcal{I}$	$\mathcal{C}$	$\mathcal{P}$	$\mathcal{T}$
$\mathcal{C}$	$\mathcal{C}$	$\mathcal{I}$	$\mathcal{T}$	$\mathcal{P}$
$\mathcal{P}$	$\mathcal{P}$	$\mathcal{T}$	$\mathcal{I}$	$\mathcal{C}$
$\mathcal{T}$	$\mathcal{T}$	$\mathcal{P}$	$\mathcal{C}$	$\mathcal{I}$

$\mathcal{C} = (C, \bar{C})$, $\mathcal{P} = (P, \bar{P})$ and $\mathcal{T} = (T, \bar{T})$, and the two elements of each coset have the same effect on the components of $\boldsymbol{e}$ and $\boldsymbol{b}$. The factor group formed by $\mathcal{I}$, $\mathcal{C}$, $\mathcal{P}$ and $\mathcal{T}$ is isomorphic to the Klein four-group; its multiplication table is displayed in Table 3. Similarly ρ and $\boldsymbol{j}$ are left invariant by the normal subgroup $(I, \bar{C})$ with cosets $(\bar{I}, C)$, (P, T) and $(\bar{P}, \bar{T})$, and w and $\boldsymbol{f}$ are left invariant by the normal subgroup (I, C) with cosets $(\bar{I}, \bar{C})$, $(P, \bar{T})$ and $(\bar{P}, T)$. It should be emphasized that the behaviour of ρ, $\boldsymbol{j}$, $\boldsymbol{e}$ and $\boldsymbol{b}$ under C, P and T is to some extent a matter of convention. The convention adopted here is such that in any frame the charge and current densities are given by Equations (2.5) and (2.6) and the Maxwell–Lorentz equations hold in the form (2.1) to (2.4). Many other conventions are possible.† On the other hand, the behaviour of w and $\boldsymbol{f}$, as exemplified by their invariance under C, is the same as for forces that are not of electromagnetic origin and is independent of the convention used for the behaviour of ρ, $\boldsymbol{j}$, $\boldsymbol{e}$ and $\boldsymbol{b}$. The charges and fields can be observed only in each other's presence and their combination in the expressions for the power and force densities is such as to ensure transformation properties for w and $\boldsymbol{f}$ that would be expected for ordinary forces.

The covariance under C, P and T that is shown by the coupled Maxwell–Lorentz and Newtonian equations is not shared by all the laws of nature. The non-conservation of parity in the weak interaction, which is responsible for the dynamics of beta emission, was suggested by Lee and Yang (1956) and subsequently confirmed experimentally. That the combined transformation of charge conjugation and parity is also not a symmetry follows from the decay (Christenson *et al.* 1964) of the long-lived neutral K meson into two charged pions, a decay that is forbidden

† Thus, e.g., if it is demanded that ρ and $\boldsymbol{j}$ together transform as a true four-vector under all Lorentz transformations, then the expressions (2.5) and (2.6) should include a negative sign for non-orthochronous reference frames, in which the sense of time is reversed (Healy 1978).

by *CP* conservation. (For reviews of this and subsequent work on *CP* asymmetry, see Fitch 1981 and Cronin 1981.) The *CPT* theorem, established on the very general assumptions of Lorentz covariance and locality (see Streater and Wightman 1964), then implies that time-reversal is also not a symmetry of the physical world. Thus the separate conservation of *C*, *P* and *T* is only an approximation which is, however, valid for phenomena that are adequately described by electrodynamics alone.

2.3 Vector and scalar potentials

The equations of motion for the fields $\boldsymbol{e}$ and $\boldsymbol{b}$ involve only first-order derivatives with respect to the time and space coordinates. In the Lagrangian formalism for continuous systems, however, the field equations contain second-order derivatives. To use this formalism for the electromagnetic field system, it is therefore necessary to recast the theory in terms of auxiliary variables for which the equations of motion are equivalent to the Maxwell–Lorentz equations but involve second-order derivatives. These auxiliary variables are the vector and scalar potentials $\boldsymbol{a}$ and ϕ. It is well known that a solenoidal vector field (the divergence of which vanishes identically) is expressible as the curl of a vector potential and that an irrotational vector field (the curl of which vanishes identically) is expressible as the negative gradient of a scalar potential. Since $\boldsymbol{b}$ is solenoidal, we must then have

$$\boldsymbol{b} = \nabla \times \boldsymbol{a} \tag{2.10}$$

for some vector field $\boldsymbol{a}$. It follows from this and Equation (2.3) that $\boldsymbol{e} + c^{-1}\,\partial \boldsymbol{a}/\partial t$ is irrotational and hence that

$$\boldsymbol{e} = -\nabla\phi - \frac{1}{c}\frac{\partial \boldsymbol{a}}{\partial t} \tag{2.11}$$

for some scalar field ϕ. We shall prove these assertions, however, by explicitly constructing line integral fields $\bar{\boldsymbol{a}}$ and $\bar{\phi}$ that satisfy Equations (2.10) and (2.11).

We suppose that $\boldsymbol{e}$ and $\boldsymbol{b}$ have continuous first derivatives in a region $\mathcal{R}$ that is deformable to a point. For each field point $\boldsymbol{x}$ in $\mathcal{R}$ we choose a fixed curve $C_{\boldsymbol{x}}$ starting at an arbitrary reference point $\boldsymbol{R}$ and ending at $\boldsymbol{x}$. The curves $C_{\boldsymbol{x}}$ are assumed to lie entirely in $\mathcal{R}$ and to be parametrized by a real variable u ranging between the limits u_1 and u_2. The parameter u can be scaled so that the limits are independent of $\boldsymbol{x}$, and this will be supposed done. The points $\boldsymbol{x}'$ of $C_{\boldsymbol{x}}$ are then specified by a function $\boldsymbol{x}'(u, \boldsymbol{x})$ such that $\boldsymbol{x}'(u_1, \boldsymbol{x}) = \boldsymbol{R}$ and $\boldsymbol{x}'(u_2, \boldsymbol{x}) = \boldsymbol{x}$. In the following we shall deal only with simple curves (which have no multiple points) consisting of a finite number of regular arcs (which have continuous tangents) and with orientable surfaces consisting of a finite number of regular segments. We shall also assume the curves $C_{\boldsymbol{x}}$ to be such that for fixed u the function $\boldsymbol{x}'(u, \boldsymbol{x})$ is continuously differentiable with respect to $\boldsymbol{x}$; neighbouring points $\boldsymbol{x}$ are then the ends of neighbouring curves $C_{\boldsymbol{x}}$. The potentials $\bar{\boldsymbol{a}}$ and $\bar{\phi}$ are now defined as line integrals along the curves $C_{\boldsymbol{x}}$ (cf. De Witt 1962, Belinfante 1962). Thus

$$\bar{a}_i(\boldsymbol{x}, t) = \int_{C_{\boldsymbol{x}}} \frac{\partial \boldsymbol{x}'}{\partial x_i} \cdot \boldsymbol{b}(\boldsymbol{x}', t) \times \mathrm{d}\boldsymbol{x}' \tag{2.12}$$

which defines $\bar{\boldsymbol{a}}$ through its components in a rectangular Cartesian coordinate system, and

$$\bar{\phi}(\boldsymbol{x}, t) = -\int_{C_{\boldsymbol{x}}} \boldsymbol{e}(\boldsymbol{x}', t) \cdot \mathrm{d}\boldsymbol{x}' \tag{2.13}$$

The values of these fields at a given place and time depend in general on the chosen reference point $\boldsymbol{R}$ and integration path $C_{\boldsymbol{x}}$. For an electrostatic field, however, it follows from application of Stokes' theorem that $\bar{\phi}$ is independent of the path if the reference point remains fixed and that changing $\boldsymbol{R}$ merely changes $\bar{\phi}$ by a constant.

To show that the line integral vector potential is indeed a solution of Equation (2.10), we consider an open surface S lying entirely in $\mathscr{R}$ and having a closed boundary curve Γ. The sense of Γ is taken to be independent of the chirality of the coordinate frame; the sense of S is determined from that of Γ by the right-hand rule in a right-handed frame and the left-hand rule in a left-handed frame. (This convention for the relation of the sense of an open surface to that of its boundary curve will be adhered to in the following. See Appendix B.) The curve Γ can be parametrized by a real variable v ranging between the limits v_1 and v_2; since Γ is closed, $\boldsymbol{x}(v_1) = \boldsymbol{x}(v_2)$. We let Σ be the surface generated by the curve $C_{\boldsymbol{x}}$ as its end point $\boldsymbol{x}$ is carried around Γ (see Fig. 4). Since $\mathscr{R}$ is deformable to a point and all the curves $C_{\boldsymbol{x}}$ (and thus also the surface $\boldsymbol{\Sigma}$) are in $\mathscr{R}$, the volume V enclosed by $\boldsymbol{\Sigma}$ and S is also in $\mathscr{R}$. The points $\boldsymbol{x}'$ of $\boldsymbol{\Sigma}$ are specified by the function $\boldsymbol{x}'\{u, \boldsymbol{x}(v)\}$ with $u = u_1$ corresponding to $\boldsymbol{R}$ and $u = u_2$ corresponding to Γ. The curve Γ will be supposed to have the same sense when considered as the boundary of Σ that it has when considered as the boundary of S. The surface element on Σ is then given by

$$\mathrm{d}\boldsymbol{S}' = \frac{\partial \boldsymbol{x}'}{\partial u} \times \frac{\partial \boldsymbol{x}'}{\partial v}\, \mathrm{d}u\, \mathrm{d}v \tag{2.14}$$

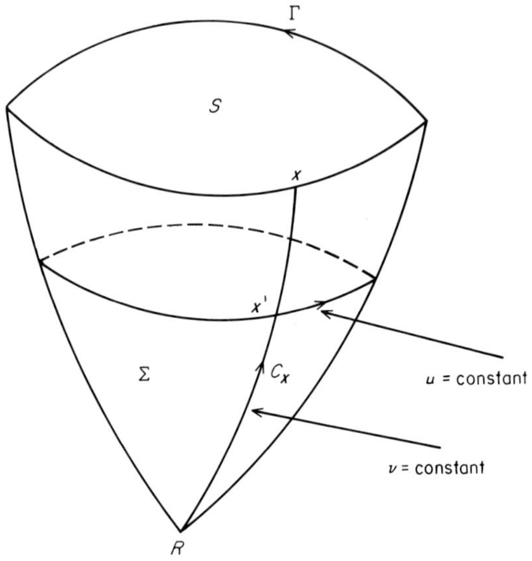

Fig. 4. Showing the surface Σ generated by the integration path $C_{\mathbf{x}}$ as its end point **x** is carried around the closed curve Γ.

Now from Stokes' theorem and Equations (2.12) and (2.14) we obtain

$$\iint_S \nabla \times \bar{\boldsymbol{a}}(\boldsymbol{x}, t) \cdot \mathrm{d}\boldsymbol{S} = \oint_\Gamma \bar{\boldsymbol{a}}(\boldsymbol{x}, t) \cdot \mathrm{d}\boldsymbol{x}$$

$$= \int_{u_1}^{u_2} \int_{v_1}^{v_2} \frac{\partial \boldsymbol{x}'}{\partial u} \times \frac{\partial \boldsymbol{x}'}{\partial x_i} \frac{\partial x_i}{\partial v} \cdot \boldsymbol{b}(\boldsymbol{x}', t) \, \mathrm{d}u \, \mathrm{d}v$$

$$= \iint_\Sigma \boldsymbol{b}(\boldsymbol{x}', t) \cdot \mathrm{d}\boldsymbol{S}'$$

$$= \iint_S \boldsymbol{b}(\boldsymbol{x}, t) \cdot \mathrm{d}\boldsymbol{S} \tag{2.15}$$

where the last step follows from the relation of the sense of Σ to that of S and from the fact that $\nabla \cdot \boldsymbol{b}$ vanishes in the volume V. Since S is an arbitrary open surface, Equation (2.10) must hold with $\boldsymbol{a} = \bar{\boldsymbol{a}}$.

To show that the line integral scalar and vector potentials together satisfy Equation (2.11), we let Γ_{12} be an arbitrary curve in $\mathcal{R}$ starting at one field point $\boldsymbol{x}_1$ and ending at another field point $\boldsymbol{x}_2$. As $\boldsymbol{x}$ is carried along Γ_{12}, the curve C generates a surface Σ_{12} which lies entirely in $\mathcal{R}$. If

$C^{(1)}$ is that curve C_x which ends at $\boldsymbol{x}_1$ and $C^{(2)}$ that which ends at $\boldsymbol{x}_2$, then the boundary curve of Σ_{12} is taken to be traced out in the positive sense by going from $\boldsymbol{R}$ to $\boldsymbol{x}_1$ along $C^{(1)}$, then from $\boldsymbol{x}_1$ to $\boldsymbol{x}_2$ along Γ_{12} and finally from $\boldsymbol{x}_2$ back to $\boldsymbol{R}$ in the negative sense along $C^{(2)}$. The parameter v will now be used for Γ_{12}, with v_1 corresponding to $\boldsymbol{x}_1$ and v_2 to $\boldsymbol{x}_2$. The surface element on Σ_{12} is then given as before by Equation (2.14). It follows from Equations (2.3), (2.12) and (2.13) that

$$\begin{aligned}
&\int_{\Gamma_{12}} \left(-\nabla\bar{\phi} - \frac{1}{c}\frac{\partial \bar{\boldsymbol{a}}}{\partial t}\right) . \mathrm{d}\boldsymbol{x} \\
&\quad = -\bar{\phi}(\boldsymbol{x}_2, t) + \bar{\phi}(\boldsymbol{x}_1, t) - \frac{1}{c}\int_{u_1}^{u_2}\int_{v_1}^{v_2} \frac{\partial \boldsymbol{x}'}{\partial u} \times \frac{\partial \boldsymbol{x}'}{\partial x_i}\frac{\partial x_i}{\partial v} . \frac{\partial}{\partial t}\boldsymbol{b}(\boldsymbol{x}', t)\,\mathrm{d}u\,\mathrm{d}v \\
&\quad = \int_{C^{(2)}} \boldsymbol{e}(\boldsymbol{x}', t) . \mathrm{d}\boldsymbol{x}' - \int_{C^{(1)}} \boldsymbol{e}(\boldsymbol{x}', t) . \mathrm{d}\boldsymbol{x}' + \iint_{\Sigma_{12}} \nabla' \times \boldsymbol{e}(\boldsymbol{x}', t) . \mathrm{d}\boldsymbol{S}' \\
&\quad = \int_{\Gamma_{12}} \boldsymbol{e}(\boldsymbol{x}, t) . \mathrm{d}\boldsymbol{x} \qquad (2.16)
\end{aligned}$$

where the last step involves use of Stokes' theorem. Since Γ_{12} is an arbitrary curve, Equation (2.11) must hold with $\phi = \bar{\phi}$ and $\boldsymbol{a} = \bar{\boldsymbol{a}}$.

PROBLEM

Show directly that the line integral vector and scalar potentials satisfy Equations (2.10) and (2.11). Differentiate with respect to $\boldsymbol{x}$ and t under the integral signs in the expressions (2.12) and (2.13), integrate by parts with respect to u, and use the source-free Maxwell–Lorentz equations.

2.4 Gauge transformations

The electromagnetic fields $\boldsymbol{e}$ and $\boldsymbol{b}$ do not uniquely determine the vector and scalar potentials $\boldsymbol{a}$ and ϕ. If we regard Equations (2.10) and (2.11) as inhomogeneous partial differential equations relating the unknown functions $\boldsymbol{a}$ and ϕ to the known source functions $\boldsymbol{e}$ and $\boldsymbol{b}$, then the line integral potentials considered in Section 2.3 are particular integrals of these equations. The general solutions are obtained by adding to the particular integrals complementary functions which are solutions of the homogeneous equations. It is easy to see that if $\boldsymbol{a}^{(1)}$ and $\phi^{(1)}$ satisfy Equations (2.10) and (2.11), then so do $\boldsymbol{a}^{(2)}$ and $\phi^{(2)}$, where

$$\boldsymbol{a}^{(2)}=\boldsymbol{a}^{(1)}-\nabla\chi \tag{2.17}$$

$$\phi^{(2)}=\phi^{(1)}+\frac{1}{c}\frac{\partial\chi}{\partial t} \tag{2.18}$$

and χ is an arbitrary differentiable function of $\boldsymbol{x}$ and t. Conversely, any two pairs of potentials satisfying Equations (2.10) and (2.11) must be related by the transformation (2.17) and (2.18) for some function χ. This is because an irrotational vector field is also lamellar (i.e. derivable from a scalar potential), so that, from Equation (2.10),

$$\boldsymbol{a}^{(2)}=\boldsymbol{a}^{(1)}-\nabla\bar{\chi} \tag{2.19}$$

for some function $\bar{\chi}(\boldsymbol{x}, t)$. Using this in conjunction with Equation (2.11) we then obtain

$$\phi^{(2)}=\phi^{(1)}+\frac{1}{c}\frac{\partial\bar{\chi}}{\partial t}+f(t) \tag{2.20}$$

where $f(t)$ is a function of t only. Equations (2.19) and (2.20) are equivalent to the transformation (2.17) and (2.18) with

$$\chi(\boldsymbol{x}, t)=\bar{\chi}(\boldsymbol{x}, t)+c\int^{t} f(t')\,\mathrm{d}t' \tag{2.21}$$

The arbitrary function χ may be restricted by the imposition of further conditions, over and above those implied by Equations (2.10) and (2.11), on $\boldsymbol{a}$ and ϕ. These further conditions are said to determine the *gauge* of the potentials. Thus the *Coulomb gauge* is defined by the subsidiary condition

$$\nabla\cdot\boldsymbol{a}=0 \tag{2.22}$$

and the *Lorentz gauge* by

$$\nabla\cdot\boldsymbol{a}+\frac{1}{c}\frac{\partial\phi}{\partial t}=0 \tag{2.23}$$

To show that it is possible to fulfil these conditions, we note that if $\boldsymbol{a}^{(1)}$ and $\phi^{(1)}$ are known potentials in some arbitrary gauge, then the Coulomb gauge condition is satisfied by $\boldsymbol{a}^{(2)}$ provided χ is a solution of the Poisson equation

$$\nabla^2\chi=\nabla\cdot\boldsymbol{a}^{(1)} \tag{2.24}$$

and the Lorentz gauge condition is satisfied by $\boldsymbol{a}^{(2)}$ and $\phi^{(2)}$ provided χ is a solution of the inhomogeneous wave equation

$$\nabla^2\chi - \frac{1}{c^2}\frac{\partial^2\chi}{\partial t^2} = \nabla \cdot \boldsymbol{a}^{(1)} + \frac{1}{c}\frac{\partial\phi^{(1)}}{\partial t} \tag{2.25}$$

The function χ that appears in the transformation (2.17) and (2.18) is known as a gauge function and the transformation itself as a gauge transformation, although transformations of this kind *within* a particular gauge may be possible for certain classes of gauge functions. Thus the Coulomb gauge condition is maintained if χ is a solution of Laplace's equation, and the Lorentz gauge condition is maintained if χ is a solution of the homogeneous wave equation.

LINE INTEGRAL POTENTIALS

By allowing the reference point $\boldsymbol{R}$ and the integration paths $C_{\boldsymbol{x}}$ to move, it is possible to introduce a class of line integral potentials wider than that discussed in Section 2.3. These more general potentials will be needed in Chapter 7 and are obtained most easily through a gauge transformation. We suppose that $\boldsymbol{R}$ traces out a smooth, though for the moment unspecified, trajectory and that the points $\boldsymbol{x}'$ of $C_{\boldsymbol{x}}$ are determined by a continuously differentiable function $\boldsymbol{x}'(u, \boldsymbol{x}, t)$ of u, $\boldsymbol{x}$ and t. The parameter u again varies between definite limits u_1 and u_2, so that $\boldsymbol{x}'(u_1, \boldsymbol{x}, t) = \boldsymbol{R}(t)$ for all $\boldsymbol{x}$ and $\boldsymbol{x}'(u_2, \boldsymbol{x}, t) = \boldsymbol{x}$ for all t. For fixed t the curves $C_{\boldsymbol{x}}$ are similar to those used previously. However, the point $\boldsymbol{x}'$ of $C_{\boldsymbol{x}}$ that corresponds to a given value of u now moves with a velocity $\partial\boldsymbol{x}'/\partial t$ which, unless $u = u_2$, need not be zero.

Let $\boldsymbol{a}$ and ϕ be potentials in some arbitrary gauge and consider the gauge transformation (to new potentials $\bar{\boldsymbol{a}}$ and $\bar{\phi}$) for which the gauge function is defined to be the integral of the old vector potential along the curve $C_{\boldsymbol{x}}$:

$$\chi(\boldsymbol{x}, t) = \int_{C_{\boldsymbol{x}}} \boldsymbol{a}(\boldsymbol{x}', t) \cdot \mathrm{d}\boldsymbol{x}' \tag{2.26}$$

Using the parametric representation of $C_{\boldsymbol{x}}$ we obtain, after integrating by parts with respect to u,

$$\nabla_i\chi = \left[a_j(\boldsymbol{x}', t)\frac{\partial x_j'}{\partial x_i}\right]_{u_1}^{u_2} + \int_{u_1}^{u_2} \frac{\partial x_j'}{\partial x_i}\left(\frac{\partial a_k}{\partial x_j'} - \frac{\partial a_j}{\partial x_k'}\right)\frac{\partial x_k'}{\partial u}\,\mathrm{d}u \tag{2.27}$$

and

$$\frac{\partial\chi}{\partial t} = [a_j(\boldsymbol{x}', t)\dot{x}_j']_{u_1}^{u_2} + \int_{u_1}^{u_2} \dot{a}_j\frac{\partial x_j'}{\partial u}\,\mathrm{d}u + \int_{u_1}^{u_2} \dot{x}_j'\left(\frac{\partial a_k}{\partial x_j'} - \frac{\partial a_j}{\partial x_k'}\right)\frac{\partial x_k'}{\partial u}\,\mathrm{d}u \tag{2.28}$$

Since $\boldsymbol{x}'=\boldsymbol{R}$ when $u=u_1$ and $\boldsymbol{x}'=\boldsymbol{x}$ when $u=u_2$, it follows from Equations (2.27) and (2.28) and Equations (2.10) and (2.11) that

$$\bar{a}_i(\boldsymbol{x}, t)=\int_{C_{\boldsymbol{x}}} \frac{\partial \boldsymbol{x}'}{\partial x_i} \cdot \boldsymbol{b}(\boldsymbol{x}', t)\times \mathrm{d}\boldsymbol{x}' \tag{2.29}$$

and

$$\bar{\phi}(\boldsymbol{x}, t)=\phi(\boldsymbol{R}, t)-\frac{1}{c}\dot{\boldsymbol{R}} \cdot \boldsymbol{a}(\boldsymbol{R}, t)-\int_{C_{\boldsymbol{x}}} \boldsymbol{e}(\boldsymbol{x}', t) \cdot \mathrm{d}\boldsymbol{x}'-\frac{1}{c}\int_{C_{\boldsymbol{x}}} \dot{\boldsymbol{x}}' \cdot \boldsymbol{b}(\boldsymbol{x}', t)\times \mathrm{d}\boldsymbol{x}' \tag{2.30}$$

These are the line integral vector and scalar potentials (De Witt 1962, Belinfante 1962, Healy 1979). It is remarkable that the vector potential depends only on the magnetic induction field $\boldsymbol{b}$ and the curves $C_{\boldsymbol{x}}$ and is the same as that given in Equation (2.12). The scalar potential (2.30), on the other hand, is independent of the old potentials $\boldsymbol{a}$ and ϕ only in special cases. This is evidently so if the reference point is always confined to a region (e.g. at infinity) in which $\boldsymbol{a}$ and ϕ vanish. Moreover, if the reference point is fixed anywhere, then $\bar{\phi}$ is independent of $\boldsymbol{a}$ and can be made independent of ϕ as well by using the modified gauge function given by

$$\chi'(\boldsymbol{x}, t)=\chi(\boldsymbol{x}, t)-c\int^{t} \phi(\boldsymbol{R}, t')\, \mathrm{d}t' \tag{2.31}$$

instead of the χ given by Equation (2.26). Both $\bar{\phi}$ and $\bar{\boldsymbol{a}}$ then depend on $\boldsymbol{e}$ and $\boldsymbol{b}$ and the curves $C_{\boldsymbol{x}}$ alone. If not only the point $\boldsymbol{R}$ but the whole curve $C_{\boldsymbol{x}}$ is fixed (for every $\boldsymbol{x}$), then the new scalar potential reduces to the simple form (2.13), provided again that χ' and not χ is used in the gauge transformation. The present treatment thus affords another proof that the potentials (2.12) and (2.13) do reproduce the fields correctly. This proof, however, in contrast to the previous one, is based on the assumption that there exist *some* valid potentials $\boldsymbol{a}$ and ϕ.

The line integral vector and scalar potentials (2.29) and (2.30) depend on the curves $C_{\boldsymbol{x}}$, and changing these curves amounts to a gauge transformation. To see this we suppose there are two reference points $\boldsymbol{R}_1$ and $\boldsymbol{R}_2$ and, for each field point $\boldsymbol{x}$, two curves $C_{\boldsymbol{x}}^{(1)}$ and $C_{\boldsymbol{x}}^{(2)}$ starting at $\boldsymbol{R}_1$ and $\boldsymbol{R}_2$, respectively, and ending at $\boldsymbol{x}$. We then have two pairs of line integral potentials, $\bar{\boldsymbol{a}}^{(1)}$ and $\bar{\phi}^{(1)}$, and $\bar{\boldsymbol{a}}^{(2)}$ and $\bar{\phi}^{(2)}$, defined as in Equations (2.29) and (2.30), and these are related to the arbitrary potentials ϕ and $\boldsymbol{a}$ through gauge transformations involving gauge functions $\chi^{(1)}$ and $\chi^{(2)}$, defined as in Equation (2.26). The potentials $\bar{\boldsymbol{a}}^{(2)}$ and $\bar{\phi}^{(2)}$ are therefore related to $\bar{\boldsymbol{a}}^{(1)}$ and $\bar{\phi}^{(1)}$ through a gauge transformation for which the gauge function $\chi^{(12)}$ is given (up to an arbitrary additive constant) by

$$\chi^{(12)}(\boldsymbol{x}, t) \equiv \chi^{(2)}(\boldsymbol{x}, t) - \chi^{(1)}(\boldsymbol{x}, t)$$
$$= \int_{C_{\boldsymbol{x}}^{(2)}} \boldsymbol{a}(\boldsymbol{x}', t) \cdot \mathrm{d}\boldsymbol{x}' - \int_{C_{\boldsymbol{x}}^{(1)}} \boldsymbol{a}(\boldsymbol{x}', t) \cdot \mathrm{d}\boldsymbol{x}' \tag{2.32}$$

We let C_{12} be a curve joining, and moving with, the points $\boldsymbol{R}_1$ and $\boldsymbol{R}_2$ and let $\Sigma_{\boldsymbol{x}}$ be a surface bounded by C_{12}, $C_{\boldsymbol{x}}^{(1)}$ (taken in its negative sense) and $C_{\boldsymbol{x}}^{(2)}$. Using Stokes' theorem and Equation (2.10) we can then express the gauge function as

$$\chi^{(12)}(\boldsymbol{x}, t) = \iint_{\Sigma_{\boldsymbol{x}}} \boldsymbol{b}(\boldsymbol{x}', t) \cdot \mathrm{d}\boldsymbol{S}' - \int_{C_{12}} \boldsymbol{a}(\boldsymbol{x}', t) \cdot \mathrm{d}\boldsymbol{x}' \tag{2.33}$$

Since $\boldsymbol{b}$ is solenoidal, the surface integral appearing here is independent of $\Sigma_{\boldsymbol{x}}$, so long as the boundary curve remains unchanged. Moreover, if $\boldsymbol{R}_1$ and $\boldsymbol{R}_2$ coincide, and C_{12} thus shrinks into a point, or if C_{12} is confined to a region in which the vector potential $\boldsymbol{a}$ vanishes, then $\chi^{(12)}$ depends only on $\boldsymbol{b}$ and the curves $C_{\boldsymbol{x}}^{(1)}$ and $C_{\boldsymbol{x}}^{(2)}$.

The gauge function (2.33) is independent of the curve C_{12} used to join $\boldsymbol{R}_1$ and $\boldsymbol{R}_2$. This is evident from Equation (2.32), as C_{12} does not appear in this equation. It follows that when $\chi^{(12)}$ is written in the form (2.33), any change in the curve C_{12} is offset by the concomitant change in the surface $\Sigma_{\boldsymbol{x}}$.

PROBLEMS

(1) Show that the Coulomb gauge condition is covariant under the subgroup of Lorentz transformations that are also Galilei transformations. (The Lorentz gauge condition, on the other hand, is covariant under the full group of inhomogeneous Lorentz transformations.)

(2) Let the relation $\sim$ be said to hold between the potentials $\boldsymbol{a}^{(1)}$ and $\phi^{(1)}$ on the one hand, and the potentials $\boldsymbol{a}^{(2)}$ and $\phi^{(2)}$ on the other, if there exists a function χ such that Equations (2.17) and (2.18) are true. Show that $\sim$ is an equivalence relation in the mathematical sense, i.e. that this relation is reflexive, symmetric and transitive.

2.5 The Coulomb gauge

The Coulomb gauge condition, namely that the vector potential be solenoidal, is the most convenient gauge condition to impose in non-relativistic theory. We first examine the relation of this gauge to certain of the line integral gauges that were discussed in Section 2.3. For this purpose it will be supposed that the region $\mathcal{R}$ consists of all space and that the reference point $\boldsymbol{R}$ is at infinity, where the fields are assumed to vanish, and that the curves $C_{\boldsymbol{x}}$ are parallel straight lines ending at the field points $\boldsymbol{x}$ and extending to infinity in a direction specified by the unit vector $\hat{\boldsymbol{\varepsilon}}$. The points $\boldsymbol{x}'$ of the line ending at $\boldsymbol{x}$ are then given by

$$\boldsymbol{x}' = \boldsymbol{x} - u\hat{\boldsymbol{\varepsilon}} \qquad -\infty < u \leqslant 0 \tag{2.34}$$

where $-u$ is the distance between $\boldsymbol{x}'$ and $\boldsymbol{x}$. The potentials associated with these straight lines are functions of $\hat{\boldsymbol{\varepsilon}}$ and may be written as

$$\boldsymbol{a}(\boldsymbol{x}, t, \hat{\boldsymbol{\varepsilon}}) = \int_{-\infty}^{0} \hat{\boldsymbol{\varepsilon}} \times \boldsymbol{b}(\boldsymbol{x}', t)\, \mathrm{d}u \tag{2.35}$$

$$\phi(\boldsymbol{x}, t, \hat{\boldsymbol{\varepsilon}}) = \int_{-\infty}^{0} \hat{\boldsymbol{\varepsilon}} \cdot \boldsymbol{e}(\boldsymbol{x}', t)\, \mathrm{d}u \tag{2.36}$$

Now the gauge of the potentials obtained from these by averaging over all directions of $\hat{\boldsymbol{\varepsilon}}$ is the Coulomb gauge (Belinfante 1962). (Note that this averaging is permissible, since the fields depend linearly on the potentials.) If $\mathrm{d}\Omega$ is an element of solid angle about $\hat{\boldsymbol{\varepsilon}}$, then the averaged vector potential may be expressed as follows:

$$\begin{aligned}\boldsymbol{a}(\boldsymbol{x}, t) &\equiv \frac{1}{4\pi} \oiint \boldsymbol{a}(\boldsymbol{x}, t, \hat{\boldsymbol{\varepsilon}})\, \mathrm{d}\Omega \\ &= \frac{1}{4\pi} \iiint \frac{(\boldsymbol{x}' - \boldsymbol{x}) \times \boldsymbol{b}(\boldsymbol{x}', t)}{|\boldsymbol{x}' - \boldsymbol{x}|^3}\, \mathrm{d}^3x' \\ &= \frac{1}{4\pi} \iiint \frac{\nabla' \times \boldsymbol{b}(\boldsymbol{x}', t)}{|\boldsymbol{x}' - \boldsymbol{x}|}\, \mathrm{d}^3x' \end{aligned} \tag{2.37}$$

where the last line results from an integration by parts and where the (positive) volume element d^3x' is $u^2\, \mathrm{d}\Omega\, \mathrm{d}(-u)$. In a similar manner we obtain

$$\begin{aligned}\phi(\boldsymbol{x}, t) &\equiv \frac{1}{4\pi} \oiint \phi(\boldsymbol{x}, t, \hat{\boldsymbol{\varepsilon}})\, \mathrm{d}\Omega \\ &= \frac{1}{4\pi} \iiint \frac{(\boldsymbol{x}' - \boldsymbol{x}) \cdot \boldsymbol{e}(\boldsymbol{x}', t)}{|\boldsymbol{x}' - \boldsymbol{x}|^3}\, \mathrm{d}^3x' \\ &= \frac{1}{4\pi} \iiint \frac{\nabla' \cdot \boldsymbol{e}(\boldsymbol{x}', t)}{|\boldsymbol{x}' - \boldsymbol{x}|}\, \mathrm{d}^3x' \end{aligned} \tag{2.38}$$

The vector potential given by Equation (2.37) is solenoidal, as may be verified by differentiating with respect to $\boldsymbol{x}$ under the integral sign. Thus *the potentials in the Coulomb gauge are the isotropic averages of the potentials in those line integral gauges for which the integration paths consist of fixed parallel straight lines coming from infinity and ending at the field points.* This conclusion is valid only if the fields drop off sufficiently rapidly at infinity for the integrals (2.35) and (2.36) to converge and for the surface terms that were omitted in Equations (2.37) and (2.38) to vanish.

The Coulomb gauge condition is maintained under gauge transformations for which the gauge function is harmonic. The potentials can be made unique, however, by stipulating that they tend to zero at least as fast as x^{-1} as x tends to infinity. For a harmonic function with space and time derivatives that are $\mathrm{O}(x^{-1})$ as x tends to infinity is at most a constant (Jeffreys and Jeffreys 1956). Now it follows from Equations (2.1), (2.5) and (2.38) that

$$\begin{aligned}\phi(\boldsymbol{x}, t) &= \iiint \frac{\rho(\boldsymbol{x}', t)}{|\boldsymbol{x}' - \boldsymbol{x}|}\,\mathrm{d}^3x' \\ &= \sum_\alpha \frac{e_\alpha}{|\boldsymbol{q}_\alpha - \boldsymbol{x}|} \\ &= \mathrm{O}(x^{-1})\end{aligned} \tag{2.39}$$

It is possible to impose on the fields boundary conditions at infinity such that $\boldsymbol{a}$ given by Equation (2.37) is also $\mathrm{O}(x^{-1})$. We henceforth assume this to be done and use the term Coulomb gauge without further qualification to refer to the potentials (2.37) and (2.38). Equation (2.39) then shows that the scalar potential is simply the instantaneous electrostatic or Coulomb potential of the system of point charges.

EQUATIONS OF MOTION IN COULOMB GAUGE

So far, only the source-free Maxwell–Lorentz equations, which are identically satisfied in any gauge, have been used. The equations of motion for the potentials are obtained by substituting the expressions (2.10) and (2.11) for $\boldsymbol{e}$ and $\boldsymbol{b}$ in the Maxwell–Lorentz equations that involve sources. In the Coulomb gauge the equations of motion are

$$\nabla \times (\nabla \times \boldsymbol{a}) + \frac{1}{c^2}\frac{\partial^2 \boldsymbol{a}}{\partial t^2} = \frac{4\pi}{c}\boldsymbol{j} - \frac{1}{c}\frac{\partial}{\partial t}\nabla\phi \tag{2.40}$$

and

$$\nabla^2\phi = -4\pi\rho \tag{2.41}$$

The scalar potential thus satisfies Poisson's equation with the charge density as source. (The solution of this equation subject to the condition that ϕ is $O(x^{-1})$ as x tends to infinity is given by Equation (2.39).) It follows from Equation (2.41) and the continuity equation that the vector field appearing on the right-hand side of Equation (2.40) is solenoidal. This may be used to express the total current density as a sum of longitudinal (i.e. irrotational) and transverse (i.e. solenoidal) parts. Thus

$$\boldsymbol{j} = \boldsymbol{j}^{\parallel} + \boldsymbol{j}^{\perp} \tag{2.42}$$

where

$$\boldsymbol{j}^{\parallel} = \frac{1}{4\pi}\frac{\partial}{\partial t}\nabla\phi \tag{2.43}$$

and

$$\nabla \times \boldsymbol{j}^{\parallel} = 0 \qquad \nabla \cdot \boldsymbol{j}^{\perp} = 0 \tag{2.44}$$

Since also the vector potential is transverse, its equation of motion can be written as

$$\nabla^2 \boldsymbol{a} - \frac{1}{c^2}\frac{\partial^2 \boldsymbol{a}}{\partial t^2} = -\frac{4\pi}{c}\boldsymbol{j}^{\perp} \tag{2.45}$$

so that in the Coulomb gauge $\boldsymbol{a}$ obeys the inhomogeneous wave equation with $\boldsymbol{j}^{\perp}$ as source. We note also that in this gauge the splitting of the electric field $\boldsymbol{e}$ into longitudinal and transverse parts is, as it were, already done, since

$$\boldsymbol{e}^{\parallel} = -\nabla\phi \qquad \boldsymbol{e}^{\perp} = -\frac{1}{c}\frac{\partial \boldsymbol{a}}{\partial t} \tag{2.46}$$

The longitudinal part $\boldsymbol{e}^{\parallel}$ is the instantaneous electrostatic field due to the charged particles, while the transverse part $\boldsymbol{e}^{\perp}$ is associated with the radiation field proper.

The division of a vector field into longitudinal and transverse components can be carried out with the help of the longitudinal and transverse delta-function dyadics $\delta^{\parallel}_{ij}(\boldsymbol{x})$ and $\delta^{\perp}_{ij}(\boldsymbol{x})$, respectively (see Appendix C). In the case of the current density we have

$$j^{\parallel}_{i}(\boldsymbol{x}, t) = \sum_{\alpha} e_{\alpha}\dot{q}_{\alpha j}\delta^{\parallel}_{ij}(\boldsymbol{x} - \boldsymbol{q}_{\alpha}) \tag{2.47}$$

$$j^{\perp}_{i}(\boldsymbol{x}, t) = \sum_{\alpha} e_{\alpha}\dot{q}_{\alpha j}\delta^{\perp}_{ij}(\boldsymbol{x} - \boldsymbol{q}_{\alpha}) \tag{2.48}$$

This partitioning of $\boldsymbol{j}$ is the same as that implied by Equations (2.42) and (2.43), as may be verified by using the expression (2.39) for ϕ.

PROBLEMS

(1) Write down the coupled equations of motion for the potentials in an arbitrary gauge. Show that in the Lorentz gauge the equations are uncoupled and that $\boldsymbol{a}$ obeys the inhomogeneous wave equation with $\boldsymbol{j}$ as source, while ϕ obeys the inhomogeneous wave equation with ρ as source.

(2) Show that the longitudinal and transverse delta-function dyadics are symmetric in i and j and are even functions of $\boldsymbol{x}$. Show also that:
(i) $\delta^{\parallel}_{ii}(\boldsymbol{x}) = \delta(\boldsymbol{x})$, $\quad \delta^{\perp}_{ii}(\boldsymbol{x}) = 2\delta(\boldsymbol{x})$
(ii) $\varepsilon_{ijk}\nabla_j\delta^{\parallel}_{kl}(\boldsymbol{x}) = 0$, $\quad \nabla_i\delta^{\perp}_{ij}(\boldsymbol{x}) = 0$
where ε is the Levi–Civita alternating pseudotensor.

2.6 Energy and momentum balance

As the charged particles are accelerated by the Lorentz force, so their energy and momentum are changed. This change can be accounted for by ascribing energy and momentum to the electromagnetic field. Let V be a fixed volume that is large enough to contain all the charged particles at some time t, and let S be its closed boundary surface. The normal to S will be supposed to point outwards from V (see Appendix B). We shall show that the change in the energy of the particles is offset by a change in the field energy contained in V together with a flux of field energy across S, and that the change in the momentum (linear or angular) of the particles is accompanied by a change in the field momentum contained in V as well as by forces or torques exerted by the field through S.

By substituting from the Maxwell–Lorentz equations on the right-hand side of the vector identity

$$\nabla \cdot (\boldsymbol{e}\times\boldsymbol{b}) = (\nabla\times\boldsymbol{e}) \cdot \boldsymbol{b} - \boldsymbol{e} \cdot (\nabla\times\boldsymbol{b}) \tag{2.49}$$

we obtain the relation

$$\nabla \cdot \boldsymbol{n} + \frac{\partial u}{\partial t} + \boldsymbol{j} \cdot \boldsymbol{e} = 0 \tag{2.50}$$

Here

$$\boldsymbol{n} = \frac{c}{4\pi}\,\boldsymbol{e}\times\boldsymbol{b} \tag{2.51}$$

and

$$u = \frac{1}{8\pi}\,(\boldsymbol{e}^2 + \boldsymbol{b}^2) \tag{2.52}$$

Equation (2.50) expresses the local form of the energy balance. For use of the divergence theorem and the Lorentz force law yields

$$\oiint_S \boldsymbol{n} \,.\, \mathrm{d}\boldsymbol{S} + \frac{\mathrm{d}U}{\mathrm{d}t} + \frac{\mathrm{d}K}{\mathrm{d}t} = 0 \tag{2.53}$$

where U is the integral of u over the volume V and K is the total kinetic energy of the charged particles. The vector $\boldsymbol{n}$, which is known as the Poynting vector, is taken to represent the flux density of the field energy. Thus $\boldsymbol{n} \,.\, \mathrm{d}\boldsymbol{S}$ is the amount of energy per unit time that crosses the surface element $\mathrm{d}S$ in the direction of $\mathrm{d}\boldsymbol{S}$. The surface integral in Equation (2.53) then gives the net outward flux through S. Similarly, the quantity u is taken to represent the energy density of the field, so that U is the field energy contained in V. Equation (2.53) is thus equivalent to the statement that the rate of increase of the total energy in V equals the rate at which energy is being carried by the field across S into V. Hence there are no sources or sinks of energy within any finite volume. There may, however, be energy escaping to, or coming from, infinity.

To investigate the momentum balance, it is convenient to introduce a second-order symmetric tensor m defined through its Cartesian components by

$$m_{ij} = \frac{1}{8\pi}(\delta_{ij}e^2 - 2e_ie_j + \delta_{ij}b^2 - 2b_ib_j) \tag{2.54}$$

This tensor is known as the Maxwell stress tensor, as the electromagnetic field can be considered as giving rise on the closed surface S to an outward force determined by

$$\mathrm{d}F_i = m_{ij}\,\mathrm{d}S_j \tag{2.55}$$

The definition (2.54) and the Maxwell–Lorentz equations together imply that

$$\frac{\partial m_{ij}}{\partial x_j} + \frac{\partial g_i}{\partial t} + f_i = 0 \tag{2.56}$$

Here $\boldsymbol{f}$ is again the Lorentz force density and

$$\mathbf{g} = \frac{1}{4\pi c}\,\boldsymbol{e} \times \boldsymbol{b} \tag{2.57}$$

The vector $\mathbf{g}$ is to be interpreted as the momentum density of the field. For on integration over the volume V, Equation (2.56) becomes

$$\oiint_S \mathrm{d}\boldsymbol{F} + \frac{\mathrm{d}\boldsymbol{G}}{\mathrm{d}t} + \frac{\mathrm{d}\boldsymbol{P}}{\mathrm{d}t} = 0 \tag{2.58}$$

where $\boldsymbol{G}$ is the integral of $\mathbf{g}$ over V and $\boldsymbol{P}$ is the total linear momentum of the charged particles. Thus the net inward force exerted by the field through S equals the rate of increase of the total linear momentum contained in V.

The balance of angular momentum in the system may be inferred from that of the linear momentum and the symmetry of the Maxwell stress tensor. It follows immediately from Equation (2.56) that

$$\frac{\partial}{\partial x_l}(\varepsilon_{ijk}x_j m_{kl})+\frac{\partial}{\partial t}(\varepsilon_{ijk}x_j g_k)+\varepsilon_{ijk}x_j f_k=0 \tag{2.59}$$

Integration of this equation over V then gives

$$\oiint_S \boldsymbol{x}\times \mathrm{d}\boldsymbol{F}+\frac{\mathrm{d}\boldsymbol{I}}{\mathrm{d}t}+\frac{\mathrm{d}\boldsymbol{M}}{\mathrm{d}t}=0 \tag{2.60}$$

where $\boldsymbol{M}$ is the total angular momentum of the charged particles about the origin (which may be an arbitrary fixed point) and $\boldsymbol{I}$ is the field angular momentum in V. Thus

$$\boldsymbol{I}=\iiint_V \boldsymbol{x}\times\mathbf{g}\,\mathrm{d}^3x \tag{2.61}$$

Now the total moment of the electromagnetic forces that act outwards through S is represented by the surface integral in Equation (2.60). Thus the sum of the moments of the inward forces equals the rate of increase of the total angular momentum in V.

TRANSFORMATION OF FIELD ENERGY AND MOMENTUM

The field energy density and energy flux density are not uniquely determined. If we carry out the transformation

$$u\to u'=u+\nabla\,.\,\boldsymbol{v} \tag{2.62}$$

$$\boldsymbol{n}\to\boldsymbol{n}'=\boldsymbol{n}-\frac{\partial\boldsymbol{v}}{\partial t} \tag{2.63}$$

with an arbitrary differentiable vector field $\boldsymbol{v}(\boldsymbol{x},t)$, then Equation (2.50) retains its form, since

$$\nabla\,.\,\mathbf{n}'+\frac{\partial u'}{\partial t}=\nabla\,.\,\boldsymbol{n}+\frac{\partial u}{\partial t} \tag{2.64}$$

Conversely, any u' and $\boldsymbol{n}'$ that satisfy Equation (2.64) must be related to u and $\boldsymbol{n}$ through the transformation (2.62) and (2.63) for some $\boldsymbol{v}$. For if $\bar{\boldsymbol{v}}$ is chosen so that $\dot{\bar{\boldsymbol{v}}}$ equals $\boldsymbol{n}-\boldsymbol{n}'$, then Equation (2.64) implies that

$$u' = u + \nabla \cdot \bar{\boldsymbol{v}} + h(\boldsymbol{x}) \tag{2.65}$$

where h is independent of t. Now let $p(\boldsymbol{x})$ be a solution of Poisson's equation with $h(\boldsymbol{x})$ as source, and put

$$\boldsymbol{v}(\boldsymbol{x}, t) = \bar{\boldsymbol{v}}(\boldsymbol{x}, t) + \nabla p(\boldsymbol{x}) \tag{2.66}$$

Then Equations (2.62) and (2.63) are satisfied with this $\boldsymbol{v}$. The field energy in V and the energy flux across S are altered by the transformation (2.62) and (2.63), but in such a way that the energy balance is preserved. The increase of amount $\oiint_S \boldsymbol{v} \cdot \mathrm{d}\boldsymbol{S}$ in the field energy is compensated for by a decrease of amount $(\mathrm{d}/\mathrm{d}t) \oiint_S \boldsymbol{v} \cdot \mathrm{d}\boldsymbol{S}$ in the outward energy flux. It is possible to take $\nabla \times \boldsymbol{v}$ to be zero without excluding any transformation that alters the energy in V or the energy flux through S, since the integral of $\boldsymbol{v}^{\perp}$ over any closed surface vanishes. We shall assume that $\nabla \times \boldsymbol{v}$ is identically zero.

The transformations of the energy density and Poynting vector that maintain the local energy balance equation form a group, and the transformations for which $\nabla \times \boldsymbol{v}$ is zero form a subgroup of this group. Similarly there is a group of transformations of the Maxwell stress tensor and the field momentum density under which Equation (2.56) is covariant. Again, however, we confine ourselves to a subgroup of the full group, namely to those transformations $m_{ij} \to m'_{ij}$ and $\mathbf{g} \to \mathbf{g}'$ that, taken in conjunction with Equations (2.62) and (2.63), preserve the following properties of u, $\boldsymbol{n}$, $\mathbf{g}$ and m:

(i) the symmetry of the Maxwell stress tensor;
(ii) the equality of the momentum density and $1/c^2$ times the Poynting vector;
(iii) the equality of the energy density and the trace of the Maxwell stress tensor.

The covariance of these properties is desirable from the point of view of the special theory of relativity.† Now property (ii) and Equation (2.63) give immediately

$$\mathbf{g}' = \mathbf{g} + \nabla s \tag{2.67}$$

† The stress–energy–momentum tensor T has components $T^{00} = u$, $T^{0i} = c^{-1} n_i$, $T^{i0} = c g_i$ and $T^{ij} = m_{ij}$. Properties (i) and (ii) imply that T is symmetric ($T^{\mu\nu} = T^{\nu\mu}$) and property (iii) that T is traceless ($T^{\mu}{}_{\mu} = 0$).

where s is a scalar function that satisfies

$$\nabla s + \frac{1}{c^2}\frac{\partial \boldsymbol{v}}{\partial t} = 0 \tag{2.68}$$

That such an s must exist is a consequence of the requirement that $\nabla \times \boldsymbol{v}$ be identically zero. It follows then that the linear momentum balance is preserved if and only if

$$m'_{ij} = m_{ij} - \delta_{ij}\frac{\partial s}{\partial t} + t_{ij} \tag{2.69}$$

where (t_{ij}) is a second-order tensor with vanishing divergence. If m' is to have property (i), (t_{ij}) must be symmetric also. These conditions on (t_{ij}) imply that the surface integrals

$$\oiint_S t_{ij}\,\mathrm{d}S_j \qquad \text{and} \qquad \oiint_S \varepsilon_{ijk}x_j t_{kl}\,\mathrm{d}S_l$$

are both zero. We could thus take (t_{ij}) to be identically zero without excluding any transformation that changes either the net force or the net torque acting on an arbitrary closed surface. (Note also that the linear and angular momentum in V are independent of (t_{ij}).) However, the transformed quantities have property (iii) only if we choose t_{ij} such that

$$t_{ii} = \nabla \cdot \boldsymbol{v} + 3\frac{\partial s}{\partial t} \tag{2.70}$$

A possible choice is

$$t_{ij} = \frac{\partial v_j}{\partial x_i} + \delta_{ij}\frac{\partial s}{\partial t} \tag{2.71}$$

with $\boldsymbol{v}$ and s being subjected to the further condition

$$\nabla \cdot \boldsymbol{v} + \frac{\partial s}{\partial t} = 0 \tag{2.72}$$

Equation (2.71) defines a symmetric tensor, since $\nabla \times \boldsymbol{v} = 0$, and the condition (2.72) ensures that this tensor is divergence free. The transformation of the stress tensor and the momentum density then becomes

$$m'_{ij} = m_{ij} + \frac{\partial v_j}{\partial x_i} \tag{2.73}$$

$$\mathbf{g}' = \mathbf{g} + \nabla s \tag{2.74}$$

Moreover, the accompanying transformation of the energy density and Poynting vector can now be written as

$$u' = u - \frac{\partial s}{\partial t} \tag{2.75}$$

$$\boldsymbol{n}' = \boldsymbol{n} - \frac{\partial \boldsymbol{v}}{\partial t} \tag{2.76}$$

Equation (2.68) and the condition that $\nabla \times \boldsymbol{v}$ be identically zero imply that there exists a scalar function ψ such that

$$s = \frac{1}{c^2}\frac{\partial \psi}{\partial t} \qquad \boldsymbol{v} = -\nabla\psi \tag{2.77}$$

The further condition (2.72) then shows that ψ must be a solution of the homogeneous wave equation. The converse of all this is also true: if ψ is any solution of the homogeneous wave equation and if s and $\boldsymbol{v}$ are defined through Equations (2.77), then the transformations (2.73) to (2.76) preserve the local energy and momentum balance equations as well as the properties (i) to (iii). Moreover, the whole formalism can be written in a relativistically covariant manner.†

Since the linear momentum balance and the symmetry of the Maxwell stress tensor are preserved under the transformation (2.73) and (2.74), the angular momentum balance is also preserved. From Equations (2.67) and (2.69), it follows that the angular momentum in V is increased by an amount $\oiint_S \boldsymbol{x} \times s \, \mathrm{d}\boldsymbol{S}$ and that the total moment of the outward forces acting on S is decreased by just the time derivative of this quantity. This corresponds to the increase of $\oiint_S s \, \mathrm{d}\boldsymbol{S}$ in the linear momentum and the decrease of $(\mathrm{d}/\mathrm{d}t) \oiint_S s \, \mathrm{d}\boldsymbol{S}$ in the net outward force.

REFERENCES

Belinfante, F. J. (1962). "Consequences of the postulate of a complete commuting set of observables in quantum electrodynamics". *Phys. Rev.* **128,** 2832.

Bergmann, P. G. (1942). "Introduction to the Theory of Relativity". Prentice-Hall, Englewood Cliffs.

Christenson, J., Cronin, J. W., Fitch, V. L. and Turlay, R. (1964). "Evidence for the 2π decay of the K_2^0 meson". *Phys. Rev. Lett.* **13,** 138.

† If we assume that ψ is a Lorentz scalar and define v by $v = (cs, \boldsymbol{v})$, then v is a four vector, since $v^\mu = \partial^\mu \psi$. The stress–energy–momentum tensor T is then transformed according to

$$T'^{\mu\nu} = T^{\mu\nu} - \partial^\mu v^\nu = T^{\mu\nu} - \partial^\mu \partial^\nu \psi$$

where $\partial_\mu \partial^\mu \psi = 0$.

Cronin, J. W. (1981). "CP symmetry violation—the search for its origin". *Rev. Mod. Phys.* **53,** 373.
De Witt, B. S. (1972) "Quantum theory without electromagnetic potentials". *Phys. Rev.* **125,** 2189.
Fitch, V. L. (1981). "Discovery of charge-conjugation parity asymmetry". *Rev. Mod. Phys.* **53,** 367.
Healy, W. P. (1978). "Covariant representation of microscopic charge and current densities in terms of polarisation and magnetisation fields". *J. Phys. A: Math. Gen.* **11,** 1899.
Healy, W. P. (1979). "Line-integral Lagrangians for the electromagnetic fields interacting with charged particles. "*Phys. Rev.* **A19,** 2353.
Jeffreys, H. and Jeffreys, B. S. (1956). "Methods of Mathematical Physics". Cambridge University Press.
Lee, T. D. and Yang, C. N. (1956). "Question of parity conservation in weak interactions." *Phys. Rev.* **104,** 254.
Lorentz, H. A. (1915). "The Theory of Electrons and its Applications to the Phenomena of Light and Radiant Heat". Teubner, Leipzig.
Maxwell, J. C. (1891). A Treatise on Electricity and Magnetism". Clarendon Press, Oxford.
Streater, R. F. and Wightman, A. S. (1964). "PCT, Spin and Statistics, and All That". Benjamin, New York.
Wigner, E. P. (1959). "Group Theory and its Applications to the Quantum Mechanics of Atomic Spectra". Academic Press, New York and London.

Chapter 3

Canonical Formalism

The quantum mechanical analogue of the classical theory of radiation that was outlined in the previous chapter may be derived by using the method of canonical quantization. This method (Dirac 1925) is an expression of Bohr's correspondence principle (Bohr 1923) and is applicable whenever a classical theory can be written in canonical form. It consists essentially of replacing the classical dynamical variables by operators in Hilbert space and the classical Poisson brackets by commutator brackets (see Fig. 5). We shall use the method to quantize the coupled system of charged particles and electromagnetic field. The appropriate canonical formalism is developed in the present chapter and the transition to the quantum theory made in Chapter 4.

3.1 Discrete systems—particles

In analytical dynamics a mechanical system having a finite, or at most a countable, number of degrees of freedom is specified when its Lagrangian function $L(q, \dot{q}, t)$ is given. The vector q defines a point in the configuration space of the system and has as its components the generalized or Lagrangian coordinates q_i. If L does not depend explicitly on t, then the system is said to be scleronomic; otherwise it is rheonomic. In many cases L is simply the sum $T+U$ of the kinetic energy T and a work function U, this latter being such that the ith component of the generalized force is given by

$$F_i = \frac{\partial U}{\partial q_i} - \frac{\mathrm{d}}{\mathrm{d}t}\frac{\partial U}{\partial \dot{q}_i} \tag{3.1}$$

The work function U may happen to be independent of the velocities, in which case it reduces to the negative of a potential energy. In the general case, however, U is velocity-dependent.

The dynamical behaviour of the system is governed by Hamilton's

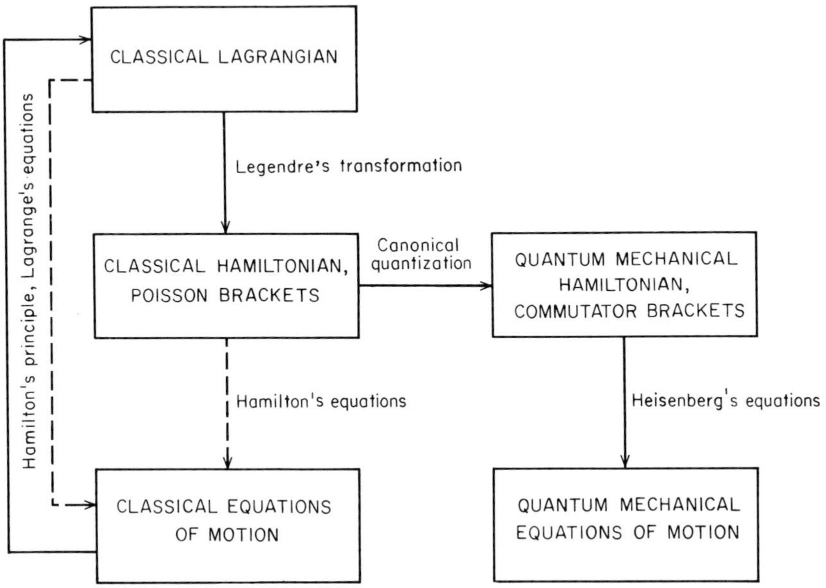

Fig. 5. Schematic representation of the method of canonical quantization.

principle: from all the kinematically possible motions that start at one configuration at time t_1 and end at another at time t_2 the natural motion is selected by the requirement that the action integral A, where

$$A = \int_{t_1}^{t_2} L(q, \dot{q}, t)\,\mathrm{d}t \tag{3.2}$$

be stationary. This means that the first-order variation δA is zero when the coordinates $q_i(t)$ of the natural trajectory are subjected to small continuously differentiable variations $\delta q_i(t)$ that vanish at t_1 and t_2 but are otherwise arbitrary. The Calculus of Variations gives the conditions for the integral (3.2) to be stationary. We have

$$\begin{aligned} \delta A &= \int_{t_1}^{t_2} \left(\frac{\partial L}{\partial q_i}\,\delta q_i + \frac{\partial L}{\partial \dot{q}_i}\,\delta \dot{q}_i \right) \mathrm{d}t \\ &= \int_{t_1}^{t_2} \left(\frac{\partial L}{\partial q_i} - \frac{\mathrm{d}}{\mathrm{d}t}\frac{\partial L}{\partial \dot{q}_i} \right) \delta q_i\,\mathrm{d}t \end{aligned} \tag{3.3}$$

as follows from an integration by parts and use of the boundary conditions.† Now each coordinate q_i is assumed to correspond to a

† Note that the time is not varied in Equation (3.3), even for a rheonomic system.

separate degree of freedom and thus each q_i can be varied independently. The variation δA is therefore zero for arbitrary small variations q_i if and only if

$$\frac{\mathrm{d}}{\mathrm{d}t}\frac{\partial L}{\partial \dot{q}_i} - \frac{\partial L}{\partial q_i} = 0 \tag{3.4}$$

for every i. These are the Euler–Lagrange equations; they are the necessary and sufficient conditions for Hamilton's principle to hold. The Lagrangian function L must be so chosen that the Euler–Lagrange equations are equivalent to the equations of motion for the system.

The Hamiltonian form of the equations of motion is obtained from the Lagrangian form by an application of Legendre's transformation. The canonical momentum p_i conjugate to the coordinate q_i is defined by

$$p_i = \frac{\partial L}{\partial \dot{q}_i} \tag{3.5}$$

and is a definite function of the coordinates and velocities (and possibly the time). It will be assumed that the relations (3.5) can be inverted to express the velocities as functions of the coordinates and momenta.† The Hamiltonian H is then given by

$$H(p, q, t) = p_i \dot{q}_i - L(q, \dot{q}, t) \tag{3.6}$$

where the velocities are to be eliminated from the right-hand side on the basis of Equations (3.5). H is thus a function of the canonical variables p_i and q_i (which together specify a point in the phase space of the system) as well as of the time. Consider now the change in H due to small variations in the coordinates and momenta only, the time remaining fixed:

$$\delta H = \frac{\partial H}{\partial p_i}\,\delta p_i + \frac{\partial H}{\partial q_i}\,\delta q_i \tag{3.7}$$

The expression (3.6), on the other hand, leads to

$$\begin{aligned} \delta H &= \dot{q}_i\,\delta p_i + p_i\,\delta \dot{q}_i - \delta L \\ &= \dot{q}_i\,\delta p_i - \dot{p}_i\,\delta q_i \end{aligned} \tag{3.8}$$

when the relations (3.5) and the Euler–Lagrange equations (3.4) have also been used. Comparing coefficients we obtain the canonical or Hamiltonian equations of motion, namely

$$\dot{q}_i = \frac{\partial H}{\partial p_i} \tag{3.9}$$

† Otherwise the Lagrangian is degenerate and the passage to the Hamiltonian requires a more general treatment (Dirac 1958).

and

$$\dot{p}_i = -\frac{\partial H}{\partial q_i} \tag{3.10}$$

The Hamiltonian equations comprise $2N$ first-order differential equations in t, if the system has N degrees of freedom; they are equivalent to the N second-order Lagrangian equations (3.4).

The *Poisson bracket* $\{f, g\}$ of two functions f and g of the canonical variables is defined by

$$\{f, g\} = \frac{\partial f}{\partial q_i}\frac{\partial g}{\partial p_i} - \frac{\partial f}{\partial p_i}\frac{\partial g}{\partial q_i} \tag{3.11}$$

It follows from the definition that if h is also a function of the canonical variables, then

$$\{f, g\} = -\{g, f\} \tag{3.12}$$

$$\{f, gh\} = \{f, g\}h + g\{f, h\} \tag{3.13}$$

and

$$\{f, \{g, h\}\} + \{g, \{h, f\}\} + \{h, \{f, g\}\} = 0 \tag{3.14}$$

The last of these relations is known as Jacobi's identity. The Poisson brackets have many remarkable properties, particularly in connection with canonical transformations (Lanczos 1970). Here we merely note that with their aid the Hamiltonian equations (3.9) and (3.10) can be written symmetrically as

$$\dot{q}_i = \{q_i, H\} \tag{3.15}$$

$$\dot{p}_i = \{p_i, H\} \tag{3.16}$$

and that, more generally, for any function f of the canonical variables (and possibly the time)

$$\dot{f} = \frac{\partial f}{\partial t} + \{f, H\} \tag{3.17}$$

This form of the equations of motion is specially suited to the quantization procedure.

PROBLEMS

(1) Show that the Lagrangian L is non-degenerate if the determinant of the Hessian matrix $(\partial^2 L/\partial \dot{q}_i\, \partial \dot{q}_j)$ is different from zero.

(2) By allowing the time to vary along with the coordinates and momenta in the derivation of the canonical equations, prove that

$$\frac{\partial H}{\partial t} = -\frac{\partial L}{\partial t}$$

and hence that for a scleronomic system the Hamiltonian, like the Lagrangian, has no explicit time dependence. Use the canonical equations to show that for such a system H is a constant of the motion.

(3) Show that the Lorentz force due to a prescribed electromagnetic field acting on a particle of charge e and position vector $\boldsymbol{q}$ may be derived from the work function:

$$U(\boldsymbol{q}, \dot{\boldsymbol{q}}, t) = \frac{e}{c}\dot{\boldsymbol{q}} \cdot \mathbf{A}(\boldsymbol{q}, t) - e\Phi(\boldsymbol{q}, t)$$

where $\mathbf{A}$ and Φ are the vector and scalar potentials in an arbitrary gauge. Hence derive the Euler–Lagrange equations corresponding to the Lagrangian

$$L(\boldsymbol{q}, \dot{\boldsymbol{q}}, t) = \tfrac{1}{2}m\dot{\boldsymbol{q}}^2 + U(\boldsymbol{q}, \dot{\boldsymbol{q}}, t)$$

and show that they are equivalent to Newton's law for a particle of mass m. Obtain the Hamiltonian for the system in the form

$$H(\boldsymbol{p}, \boldsymbol{q}, t) = \frac{1}{2m}\left\{\boldsymbol{p} - \frac{e}{c}\mathbf{A}(\boldsymbol{q}, t)\right\}^2 + e\Phi(\boldsymbol{q}, t)$$

and verify that the canonical equations also lead to Newton's law.

(4) Using just the fundamental Poisson bracket relations $\{q_i, q_j\} = 0$, $\{p_i, p_j\} = 0$ and $\{q_i, p_j\} = \delta_{ij}$, but assuming the algebraic properties (3.12) and (3.13), prove that if f has a power series expansion, then

$$\{f, p_i\} = \frac{\partial f}{\partial q_i} \qquad \{f, q_i\} = -\frac{\partial f}{\partial p_i}$$

Hence deduce the expression for the Poisson bracket of any two functions f and g that have power series expansions.

(5) The *canonical integral* is the action integral with the Lagrangian expressed as a function of p_i, q_i and their time derivatives:

$$A = \int_{t_1}^{t_2} [p_i\dot{q}_i - H(p, q, t)]\,dt$$

Show that if in the application of Hamilton's principle to this integral the q_i and p_i are treated as Lagrangian "coordinates" and the $\dot{q}_i$ and $\dot{p}_i$ as the corresponding "velocities", then the Euler–Lagrange equations are equivalent to the canonical equations of motion. This result is due to the special form of the canonical integral, in which the "kinetic energy" term is just $p_i\dot{q}_i$.

3.2 Continuous systems—fields

FUNCTIONALS AND FUNCTIONAL DERIVATIVES

Certain dynamical systems have an uncountable number of degrees of freedom. A field a (with M components a_i) is such a system. To specify the "configuration" of the field at any time t, the values of the "coordinates" $a_i(\boldsymbol{x}, t)$ must be given for all $\boldsymbol{x}$ ($\boldsymbol{x} \in R^3$) as well as for all i ($i = 1, 2, \ldots, M$). The canonical formalism for continuous systems can be developed along the same lines as that for discrete systems. Instead of dealing with functions $f(q)$ of the generalized coordinates, however, we now have to deal with functionals $F[a]$ of the field (Volterra 1959). Just as f assigns a number to the vector q, so F assigns a number to the field a, i.e. the value of the functional is determined by the values of the field components throughout all space. The functional F may be explicitly time-dependent, since it may change with time even when the field a does not; in this case it is denoted by $F[a; t]$. We shall be interested mainly in functionals that can be written as the integral over all space of a functional density $\mathscr{F}$, so that

$$F[a; t] = \iiint \mathscr{F}(a, \nabla a, t)\, \mathrm{d}^3x \tag{3.18}$$

where $\mathscr{F}$ is a function of the field components a_i, their first spatial derivatives $\nabla_j a_i$ (or $a_{i,j}$) and possibly the time t. Higher spatial derivatives could also be included, but this generalization will not be needed here.

The variation of F that is induced by small variations in the field components a_i only (the time t not being varied) is given by

$$\delta F = \iiint \left(\frac{\partial \mathscr{F}}{\partial a_i}\, \delta a_i + \frac{\partial \mathscr{F}}{\partial a_{i,j}}\, \delta a_{i,j} \right) \mathrm{d}^3x \tag{3.19}$$

It will be assumed that as x tends to infinity, the field tends to zero so rapidly that integration by parts and omission of boundary terms having the form of surface integrals at infinity is always permissible. This will be taken to apply to the variations of the field as well. If we carry out an integration by parts and put

$$\frac{\delta F}{\delta a_i} = \frac{\partial \mathscr{F}}{\partial a_i} - \frac{\partial}{\partial x_j}\frac{\partial \mathscr{F}}{\partial a_{i,j}} \tag{3.20}$$

then Equation (3.19) becomes

$$\delta F = \iiint \frac{\delta F}{\delta a_i}\,\delta a_i\,\mathrm{d}^3x \tag{3.21}$$

$\delta F/\delta a_i$ may be regarded as a kind of derivative of F with respect to that field "coordinate" which is labelled by i and $\boldsymbol{x}$. It is known as a *functional derivative.* For an ordinary function f of the generalized coordinates q_i we have the chain rule

$$\delta f = \frac{\partial f}{\partial q_i}\,\delta q_i \tag{3.22}$$

Equation (3.21) is the counterpart of this for functionals, the partial derivatives being replaced by functional derivatives and the sum over the generalized coordinates being replaced by a sum over the field components together with an integration over all space.

LAGRANGIAN AND HAMILTONIAN

The Lagrangian for a continuous system is taken to be a functional $L[a, \dot{a}; t]$ of the field and its time derivative, i.e. L is determined at any instant when both a and $\dot{a}$ are known functions of $\boldsymbol{x}$. The variation of L for fixed t is then

$$\delta L = \iiint \left(\frac{\delta L}{\delta a_i}\,\delta a_i + \frac{\delta L}{\delta \dot{a}_i}\,\delta \dot{a}_i\right)\mathrm{d}^3x \tag{3.23}$$

This follows directly from the formula (3.21), if a and $\dot{a}$ together are temporarily regarded as a single field with $2M$ independent components. Hamilton's principle now states that

$$\delta \int_{t_1}^{t_2} L[a, \dot{a}; t]\,\mathrm{d}t = 0 \tag{3.24}$$

for arbitrary small variations δa_i such that $\delta a_i(\boldsymbol{x}, t_1) \equiv 0 \equiv \delta a_i(\boldsymbol{x}, t_2)$. Using Equation (3.23), integrating by parts with respect to t and applying the boundary conditions, we obtain

$$\int_{t_1}^{t_2} \iiint \left(\frac{\delta L}{\delta a_i} - \frac{\partial}{\partial t}\frac{\delta L}{\delta \dot{a}_i}\right)\delta a_i\,\mathrm{d}^3x\,\mathrm{d}t = 0 \tag{3.25}$$

Since the variations δa_i are arbitrary, the integrand must vanish at every point and hence

$$\frac{\partial}{\partial t}\frac{\delta L}{\delta \dot{a}_i}-\frac{\delta L}{\delta a_i}=0 \tag{3.26}$$

These are the Euler–Lagrange field equations; they are the analogues of Equations (3.4) for particles. Just as for a discrete system, the Lagrangian for a continuous system must be so chosen that the Euler–Lagrange equations are equivalent to the equations of motion.

It is assumed that the functional density $\mathscr{L}$ that corresponds to the Lagrangian does not depend on the spatial derivatives of $\dot{a}$, although it may depend on those of a. The field momentum π_i canonically conjugate to a_i is then also independent of $\nabla \dot{a}$, since, by definition,

$$\pi_i(\boldsymbol{x}, t)=\frac{\delta L}{\delta \dot{a}_i}=\frac{\partial \mathscr{L}}{\partial \dot{a}_i} \tag{3.27}$$

We shall again deal only with non-degenerate Lagrangians, for which Equations (3.27) can be solved to express the velocities $\dot{a}_i$ in terms of the momenta π_i, the coordinates a_i and their spatial derivatives $a_{i,j}$. The Hamiltonian for the system, defined by

$$H[\pi, a; t]=\iiint \pi_i \dot{a}_i \, \mathrm{d}^3x - L \tag{3.28}$$

is thus a functional of π and a. The corresponding Hamiltonian density is given by

$$\mathscr{H}(\pi, a, \nabla a, t)=\pi_i \dot{a}_i-\mathscr{L}(a, \nabla a, \dot{a}, t) \tag{3.29}$$

the velocities being eliminated from the right-hand side through use of Equations (3.27). Now the variation of H at fixed t can be written in two ways. It follows from the properties of the functional derivatives that

$$\delta H=\iiint\left(\frac{\delta H}{\delta \pi_i}\,\delta \pi_i+\frac{\delta H}{\delta a_i}\,\delta a_i\right) \mathrm{d}^3x \tag{3.30}$$

and from Equations (3.26) to (3.29) that

$$\delta H=\iiint (\dot{a}_i\, \delta \pi_i-\dot{\pi}_i\, \delta a_i)\, \mathrm{d}^3x \tag{3.31}$$

Equating coefficients in these two expressions gives the canonical equations for the field, namely

$$\dot{a}_i=\frac{\delta H}{\delta \pi_i} \tag{3.32}$$

and

$$\dot{\pi}_i = -\frac{\delta H}{\delta a_i} \tag{3.33}$$

Again the Hamiltonian equations contain twice as many variables as the Lagrangian equations but are first-order in the time.

The concept of Poisson brackets can be extended to continuous systems by a natural generalization of the definition (3.11). If F and G are two functionals of the canonical variables, their Poisson bracket is taken to be

$$\{F, G\} = \iiint \left(\frac{\delta F}{\delta a_i} \frac{\delta G}{\delta \pi_i} - \frac{\delta F}{\delta \pi_i} \frac{\delta G}{\delta a_i} \right) \mathrm{d}^3 x \tag{3.34}$$

These brackets have the algebraic properties (3.12) to (3.14). Moreover, the canonical equations may be written in terms of them as

$$\dot{a}_i = \{a_i, H\} \tag{3.35}$$

$$\dot{\pi}_i = \{\pi_i, H\} \tag{3.36}$$

For this purpose it is necessary to regard $a_i(\boldsymbol{x}, t)$ and $\pi_i(\boldsymbol{x}, t)$ as functionals, for every $\boldsymbol{x}$ and t. This is possible, since the field and its conjugate momentum can be formally expressed as integrals over functional densities. Thus

$$a_i(\boldsymbol{x}, t) = \iiint \delta(\boldsymbol{x} - \boldsymbol{x}') a_i(\boldsymbol{x}', t)\, \mathrm{d}^3 x' \tag{3.37}$$

and similarly for $\pi_i(\boldsymbol{x}, t)$. From these representations and the definition (3.34) it follows that

$$\{a_i(\boldsymbol{x}, t), a_j(\boldsymbol{x}', t)\} = 0 \tag{3.38}$$

$$\{\pi_i(\boldsymbol{x}, t), \pi_j(\boldsymbol{x}', t)\} = 0 \tag{3.39}$$

and

$$\{a_i(\boldsymbol{x}, t), \pi_j(\boldsymbol{x}', t)\} = \delta_{ij}\, \delta(\boldsymbol{x} - \boldsymbol{x}') \tag{3.40}$$

These are the fundamental Poisson bracket relations for the canonical variables that describe the field.

PROBLEMS

(1) A functional does not uniquely determine its functional density, since the divergence of a vector function of the field can always be added to the latter. Verify that the functional derivative is unaltered by the addition of such a divergence.

(2) The product FG of two functionals F and G is defined by $(FG)[a]=F[a]G[a]$. Show that if F and G are expressible as integrals over functional densities, then

$$\delta(FG)=\iiint \frac{\delta(FG)}{\delta a_i}\,\delta a_i\,\mathrm{d}^3x$$

where

$$\frac{\delta(FG)}{\delta a_i}=F\frac{\delta G}{\delta a_i}+\frac{\delta F}{\delta a_i}G$$

Hence verify the algebraic properties of Poisson brackets for such functionals.

(3) By allowing the time to vary along with the field and its conjugate momentum in the derivation of the canonical equations, prove that

$$\iiint \frac{\partial \mathcal{H}}{\partial t}\,\mathrm{d}^3x=-\iiint \frac{\partial \mathcal{L}}{\partial t}\,\mathrm{d}^3x$$

Show further that this equation is unaffected by adding to $\mathcal{L}$ or $\mathcal{H}$ the divergence of an explicitly time-dependent vector function of the field.

(4) Write down the Euler–Lagrange equations corresponding to the Lagrangian density

$$\mathcal{L}(a,\nabla a,\dot{a})=\frac{1}{2}\left(\frac{\partial a_i}{\partial t}\frac{\partial a_i}{\partial t}-c^2\frac{\partial a_i}{\partial x_j}\frac{\partial a_i}{\partial x_j}\right)$$

and show that they lead to the three-dimensional wave equation

$$\left(\nabla^2-\frac{1}{c^2}\frac{\partial^2}{\partial t^2}\right)a=0$$

Show also that the Lagrangian is non-degenerate, and obtain the canonical equations for the system.

3.3 Transverse fields

In this section we consider how the canonical formalism is to be modified when the field is a three-dimensional vector field $\boldsymbol{a}$ that is subject to the condition $\nabla\cdot\boldsymbol{a}=0$. The modified formalism is directly applicable to electrodynamics in the Coulomb gauge.

The Lagrangian is now a functional $L[\boldsymbol{a},\dot{\boldsymbol{a}};t]$ of the transverse fields $\boldsymbol{a}$

and $\dot{\boldsymbol{a}}$. (Note that if $\boldsymbol{a}$ is transverse for all t, then so is $\dot{\boldsymbol{a}}$.) The condition $\nabla \cdot \boldsymbol{a}=0$ will be regarded as a kinematical constraint that holds for the varied as well as for the natural motion. Hamilton's principle is then

$$\delta \int_{t_1}^{t_2} L \,\mathrm{d}t = 0 \tag{3.41}$$

for arbitrary small *transverse* variations $\delta \boldsymbol{a}^{\perp}$ that vanish everywhere at t_1 and t_2. Variational problems with constraints are usually dealt with by the method of Lagrangian multipliers (Lanczos 1970). Here an entirely equivalent but more direct approach will be used.

If the variations are restricted by the transversality condition, then Hamilton's principle does not yield the usual form of the Euler–Lagrange equations. For the action integral is stationary if and only if

$$\int_{t_1}^{t_2} \iiint \left(\frac{\delta L}{\delta a_i} - \frac{\partial}{\partial t} \frac{\delta L}{\delta \dot{a}_i} \right) \delta a_i^{\perp} \,\mathrm{d}^3 x \,\mathrm{d}t = 0 \tag{3.42}$$

where now the variations are transverse and i runs from 1 to 3. Let us put

$$I_i(\boldsymbol{x}, t) = \frac{\delta L}{\delta a_i} - \frac{\partial}{\partial t} \frac{\delta L}{\delta \dot{a}_i} \tag{3.43}$$

and assume that I_i is continuous and $\delta a_i^{\perp}$ continuously differentiable. It cannot be concluded that I_i is identically zero, since the longitudinal part of $\boldsymbol{I}$ is completely undetermined by Equation (3.42). This follows at once from the theorem (which itself follows from the divergence theorem) that the integral over all space of the scalar product of a longitudinal and transverse field vanishes, if the fields drop off sufficiently rapidly at infinity. Two further applications of the same theorem, however, show that the transverse part of $\boldsymbol{I}$ must be zero. We have

$$\begin{aligned} 0 &= \int_{t_1}^{t_2} \iiint \boldsymbol{I} \cdot \delta \boldsymbol{a}^{\perp} \,\mathrm{d}^3 x \,\mathrm{d}t \\ &= \int_{t_1}^{t_2} \iiint \boldsymbol{I}^{\perp} \cdot \delta \boldsymbol{a}^{\perp} \,\mathrm{d}^3 x \,\mathrm{d}t \\ &= \int_{t_1}^{t_2} \iiint \boldsymbol{I}^{\perp} \cdot \delta \boldsymbol{a} \,\mathrm{d}^3 x \,\mathrm{d}t \end{aligned} \tag{3.44}$$

Since $\delta \boldsymbol{a}$ may be an arbitrary small variation (not necessarily transverse), we conclude that $\boldsymbol{I}^{\perp}$ is identically zero. The restricted variational principle thus leads to the *transverse* Euler–Lagrange equations

$$\frac{\partial}{\partial t} \frac{\delta L}{\delta \dot{a}_i^{\perp}} - \frac{\delta L}{\delta a_i^{\perp}} = 0 \tag{3.45}$$

where $\delta L/\delta \dot{a}_i^\perp$ and $\delta L/\delta a_i^\perp$ are transverse functional derivatives,† i.e., they are the ith components of the transverse parts of the vectors whose ith components are the ordinary functional derivatives $\delta L/\delta \dot{a}_i$ and $\delta L/\delta a_i$, respectively. It may be shown in a similar manner that if the condition $\nabla \times \boldsymbol{a} = 0$ is imposed as a kinematical constraint, then Hamilton's principle gives only the longitudinal part of the Euler–Lagrange equations.

We can set up a Hamiltonian formalism for transverse fields by first defining the field momentum canonically conjugate to a_i through the equation

$$\pi_i = \frac{\delta L}{\delta \dot{a}_i^\perp} \tag{3.46}$$

so that $\boldsymbol{\pi}$, like $\boldsymbol{a}$, is transverse. The Lagrangian is again assumed to be non-degenerate, and the Hamiltonian is a functional of the transverse fields $\boldsymbol{\pi}$ and $\boldsymbol{a}$,

$$H[\boldsymbol{\pi}, \boldsymbol{a}; t] = \iiint \pi_i \dot{a}_i \, \mathrm{d}^3x - L \tag{3.47}$$

The variation of H due to small transverse variations in $\boldsymbol{\pi}$ and $\boldsymbol{a}$ is then given equally well by

$$\delta H = \iiint \left(\frac{\delta H}{\delta \pi_i^\perp} \, \delta \pi_i^\perp + \frac{\delta H}{\delta a_i^\perp} \, \delta a_i^\perp \right) \mathrm{d}^3x \tag{3.48}$$

or by

$$\delta H = \iiint (\dot{a}_i \, \delta \pi_i^\perp - \dot{\pi}_i \, \delta a_i^\perp) \, \mathrm{d}^3x \tag{3.49}$$

As the coefficients of the variations in these two expressions are the components of transverse vectors (and it is for this reason that π_i in Equation (3.46) was defined to be a transverse rather than an ordinary functional derivative), the transverse variations may be replaced by any small variations whatsoever [cf. Equation (3.44)]. The coefficients can thus be equated to give the transverse Hamiltonian equations

$$\dot{a}_i = \frac{\delta H}{\delta \pi_i^\perp} \tag{3.50}$$

† With this notation the variation of a functional $F[\boldsymbol{a}]$ due to just transverse variations of $\boldsymbol{a}$ may be written as

$$\delta F = \iiint \frac{\delta F}{\delta a_i^\perp} \, \delta a_i^\perp \, \mathrm{d}^3x$$

and

$$\dot{\pi}_i = -\frac{\delta H}{\delta a_i^{\perp}} \tag{3.51}$$

To write the Hamiltonian equations in Poisson bracket form, the definition (3.34) must be modified in such a way that the Poisson brackets of $\boldsymbol{a}$ and H and of $\boldsymbol{\pi}$ and H are automatically transverse. This may be done by taking the Poisson bracket of two functionals F and G of the transverse canonical variables to be

$$\{F, G\} = \iiint \left(\frac{\delta F}{\delta a_i^{\perp}} \frac{\delta G}{\delta \pi_i^{\perp}} - \frac{\delta F}{\delta \pi_i^{\perp}} \frac{\delta G}{\delta a_i^{\perp}} \right) \mathrm{d}^3 x \tag{3.52}$$

These brackets retain the algebraic properties of the ordinary Poisson brackets. The canonical equations (3.50) and (3.51) now become

$$\dot{a}_i = \{a_i, H\} \tag{3.53}$$

and

$$\dot{\pi}_i = \{\pi_i, H\} \tag{3.54}$$

Here again a_i and π_i have been regarded as functionals for every $\boldsymbol{x}$ and t. Since $\boldsymbol{a}$ and $\boldsymbol{\pi}$ are transverse, we have

$$a_i(\boldsymbol{x}, t) = \iiint \delta_{ij}^{\perp}(\boldsymbol{x} - \boldsymbol{x}') a_j(\boldsymbol{x}', t) \, \mathrm{d}^3 x' \tag{3.55}$$

and similarly for π_i. Accordingly, the fundamental Poisson brackets are

$$\{a_i(\boldsymbol{x}, t), a_j(\boldsymbol{x}', t)\} = 0 \tag{3.56}$$

$$\{\pi_i(\boldsymbol{x}, t), \pi_j(\boldsymbol{x}', t)\} = 0 \tag{3.57}$$

and

$$\{a_i(\boldsymbol{x}, t), \pi_j(\boldsymbol{x}', t)\} = \delta_{ij}^{\perp}(\boldsymbol{x} - \boldsymbol{x}') \tag{3.58}$$

The occurrence of the transverse rather than the ordinary delta function on the right-hand side of Equation (3.58) will be seen to be of crucial importance when the quantum conditions are imposed on the electromagnetic field in the Coulomb gauge.

3.4 Canonical formulation of electrodynamics in the Coulomb gauge

The interacting system of electromagnetic field and non-relativistic charged particles does not fall completely into either of the two categories discussed previously, since this system is specified partly by a discrete set of variables, namely the coordinates of the charged particles, and partly by a continuous set, which we take to be the values of the vector potential in the Coulomb gauge. The Lagrangian L will thus be a functional of $\boldsymbol{a}$ and $\dot{\boldsymbol{a}}$ if the particle coordinates and velocities are fixed, and a function of the $\boldsymbol{q}_\alpha$ and $\dot{\boldsymbol{q}}_\alpha$ if the vector potential and its time derivative are fixed. We write $L = L[\boldsymbol{a}, \dot{\boldsymbol{a}}; q, \dot{q})$. In the application of Hamilton's principle, the particle and field coordinates are to be varied independently. The Lagrangian must then be chosen so that variation with respect to the particle coordinates gives Newton's law (2.8) and variation with respect to the field coordinates (subject to the Coulomb gauge condition) gives the equation of motion (2.45) for the vector potential. This latter equation leads only to the transverse part of the Maxwell–Lorentz equation (2.4) and the remaining Maxwell–Lorentz equations are not derived from the variational principle. The source-free equations (2.2) and (2.3) are already implicit in the relation of $\boldsymbol{e}$ and $\boldsymbol{b}$ to the potentials, while Equation (2.1) and the longitudinal part of Equation (2.4) follow from the expression (2.39) for the scalar potential ϕ. Thus ϕ is to be regarded not as a dynamical variable of the field, but as a prescribed function of the particle coordinates.

A suitable Lagrangian is obtained by setting

$$L = L_{\text{par}} + L_{\text{rad}} + L_{\text{int}} \tag{3.59}$$

with

$$L_{\text{par}} = \frac{1}{2}\sum_\alpha m_\alpha \dot{\boldsymbol{q}}_\alpha^2 - \frac{1}{2}\sum_{\alpha\neq\beta} \frac{e_\alpha e_\beta}{|\boldsymbol{q}_\alpha - \boldsymbol{q}_\beta|} \tag{3.60}$$

$$L_{\text{rad}} = \frac{1}{8\pi}\iiint \left\{\frac{1}{c^2}\dot{\boldsymbol{a}}^2 - (\nabla\times\boldsymbol{a})^2\right\} \mathrm{d}^3x \tag{3.61}$$

and

$$L_{\text{int}} = \frac{1}{c}\iiint \boldsymbol{j}\cdot\boldsymbol{a}\,\mathrm{d}^3x$$

$$= \sum_\alpha \frac{e_\alpha}{c}\dot{\boldsymbol{q}}_\alpha\cdot\boldsymbol{a}(\boldsymbol{q}_\alpha, t) \tag{3.62}$$

L_{par} is the Lagrangian appropriate to a system of charged particles interacting solely through instantaneous Coulomb forces; it has the simple form of "kinetic energy minus potential energy". L_{rad} is the Lagrangian for a radiation field far removed from charges and currents, and has the form of "electric field energy minus magnetic induction field energy". The interaction Lagrangian L_{int} couples the particle variables to the field variables. The part of the Lagrangian that involves the vector potential can be written as the integral over a Lagrangian density $\mathscr{L}$,

$$\mathscr{L} = \frac{1}{8\pi}\left\{\frac{1}{c^2}\dot{\boldsymbol{a}}^2 - (\nabla \times \boldsymbol{a})^2\right\} + \frac{1}{c}\,\boldsymbol{j}\,.\,\boldsymbol{a} \tag{3.63}$$

Since the Coulomb gauge condition is being used as a constraint, the transverse current density $\boldsymbol{j}^{\perp}$ could be substituted for the total current density $\boldsymbol{j}$ without affecting the Lagrangian. The Lagrangian density $\mathscr{L}$, on the other hand, would be altered by the addition of a divergence. It is easy to see, however, that the transverse functional derivatives remain unchanged by this addition.

To verify that L given by Equation (3.59) is indeed a suitable Lagrangian, we write down the Euler–Lagrange equations, beginning with those for the particle coordinates. We have

$$\begin{aligned} 0 &= \frac{\mathrm{d}}{\mathrm{d}t}\frac{\partial L}{\partial \dot{q}_{\alpha i}} - \frac{\partial L}{\partial q_{\alpha i}} \\ &= m_\alpha \ddot{q}_{\alpha i} - e_\alpha \left\{ \sum_{\beta \neq \alpha} \frac{e_\beta}{|\boldsymbol{q}_\alpha - \boldsymbol{q}_\beta|^3}(q_{\alpha i} - q_{\beta i}) - \frac{1}{c}\frac{\partial a_i}{\partial t} \right\} \\ &\quad - \frac{e_\alpha}{c}\dot{q}_{\alpha j}\left(\frac{\partial a_j}{\partial q_{\alpha i}} - \frac{\partial a_i}{\partial q_{\alpha j}}\right) \end{aligned} \tag{3.64}$$

The first term in the curly brackets is the longitudinal electric field $e_i^{\parallel}(\boldsymbol{x}, t)$ evaluated at $\boldsymbol{x} = \boldsymbol{q}_\alpha$. The contribution to $e_i^{\parallel}$ from the field of particle α, however, is missing. Because of the assumed point structure of the particles, this field is infinite at the location of particle α itself, and the static field energy associated with it diverges. The infinite terms have been specifically excluded from the potential energy in Equation (3.60), so that the longitudinal field acting on any particle is the combined field due to the other particles only. Returning to Equation (3.64) and using the expressions (2.10) and (2.46) for the transverse fields in terms of the vector potential, we find that

$$m_\alpha \ddot{\boldsymbol{q}}_\alpha = e_\alpha \boldsymbol{e}(\boldsymbol{q}_\alpha, t) + \frac{e_\alpha}{c}\dot{\boldsymbol{q}}_\alpha \times \boldsymbol{b}(\boldsymbol{q}_\alpha, t) \tag{3.65}$$

which is just Newton's law with the Lorentz force. To obtain the field equations, the functional derivatives of L must first be calculated from the Lagrangian density (3.63). Now

$$\begin{aligned} \frac{\delta L}{\delta a_i} &= \frac{\partial \mathscr{L}}{\partial a_i} - \frac{\partial}{\partial x_j}\frac{\partial \mathscr{L}}{\partial a_{i,j}} \\ &= \frac{1}{c}j_i - \frac{1}{4\pi}\nabla \times (\nabla \times \boldsymbol{a})|_i \end{aligned} \tag{3.66}$$

Since $\nabla\times(\nabla\times\boldsymbol{a})$ is transverse, it follows that

$$\frac{\delta L}{\delta a_i^{\perp}}=\frac{1}{c}j_i^{\perp}-\frac{1}{4\pi}\nabla\times(\nabla\times\boldsymbol{a})|_i \tag{3.67}$$

We have also

$$\begin{aligned}\frac{\delta L}{\delta \dot{a}_i}=\frac{\partial\mathscr{L}}{\partial\dot{a}_i}&=\frac{1}{4\pi c^2}\dot{a}_i\\ &=\frac{\delta L}{\delta\dot{a}_i^{\perp}}\end{aligned} \tag{3.68}$$

since $\dot{\boldsymbol{a}}$ is transverse. The Euler–Lagrange field equations (3.45) are therefore equivalent to

$$\nabla\times(\nabla\times\boldsymbol{a})+\frac{1}{c^2}\ddot{\boldsymbol{a}}=\frac{4\pi}{c}\boldsymbol{j}^{\perp} \tag{3.69}$$

which is the expected equation of motion for the vector potential in the Coulomb gauge.

The canonical momenta associated with the Lagrangian (3.59) are defined in the usual way as

$$p_{\alpha i}=\frac{\partial L}{\partial\dot{q}_{\alpha i}}=m_\alpha\dot{q}_{\alpha i}+\frac{e_\alpha}{c}a_i(\boldsymbol{q}_\alpha, t) \tag{3.70}$$

and

$$\pi_i(\boldsymbol{x}, t)=\frac{\delta L}{\delta\dot{a}_i^{\perp}}=\frac{1}{4\pi c^2}\dot{a}_i(\boldsymbol{x}, t) \tag{3.71}$$

It should be noted that the field momentum $\boldsymbol{\pi}$ is essentially the transverse electric field $\boldsymbol{e}^{\perp}$, and that in the presence of the electromagnetic field the canonical momentum $\boldsymbol{p}_\alpha$ of particle α differs from its physical momentum $m_\alpha\dot{\boldsymbol{q}}_\alpha$. Since the relations (3.70) and (3.71) can be inverted to express the velocities as functions of the coordinates and momenta, the Lagrangian is non-degenerate. We can therefore pass immediately, by means of Legendre's transformation, to the canonical formalism. The Hamiltonian is a functional of the transverse fields $\boldsymbol{\pi}$ and $\boldsymbol{a}$ and a function of the $\boldsymbol{p}_\alpha$ and $\boldsymbol{q}_\alpha$. We have

$$\begin{aligned}H[\boldsymbol{\pi},\boldsymbol{a};p,q)&=\sum_\alpha\boldsymbol{p}_\alpha\cdot\dot{\boldsymbol{q}}_\alpha+\iiint\boldsymbol{\pi}\cdot\dot{\boldsymbol{a}}\,\mathrm{d}^3x-L\\ &=\sum_\alpha\frac{1}{2m_\alpha}\left\{\boldsymbol{p}_\alpha-\frac{e_\alpha}{c}\boldsymbol{a}(\boldsymbol{q}_\alpha, t)\right\}^2+\frac{1}{2}\sum_{\alpha\neq\beta}\frac{e_\alpha e_\beta}{|\boldsymbol{q}_\alpha-\boldsymbol{q}_\beta|}\\ &\quad+\frac{1}{8\pi}\iiint\{(4\pi c\boldsymbol{\pi})^2+(\nabla\times\boldsymbol{a})^2\}\,\mathrm{d}^3x\end{aligned} \tag{3.72}$$

The canonical equations for the coupled system are deduced by a combination of the methods used previously. The particle equations are derived by keeping the field variables fixed (i.e. by choosing their variations to be zero) and repeating the analysis of Section 3.1, while the field equations are obtained by keeping the particle variables fixed and following the analysis of Section 3.3. The Hamiltonian density is given by

$$\mathcal{H} = \frac{1}{8\pi}\{(4\pi c\boldsymbol{\pi})^2 + (\nabla \times \boldsymbol{a})^2\} - \sum_\alpha \frac{e_\alpha}{m_\alpha c}\delta(\boldsymbol{x}-\boldsymbol{q}_\alpha)\boldsymbol{p}_\alpha \cdot \boldsymbol{a} + \sum_\alpha \frac{e_\alpha^2}{2m_\alpha c^2}\delta(\boldsymbol{x}-\boldsymbol{q}_\alpha)\boldsymbol{a}^2 \tag{3.73}$$

The Hamiltonian equations reproduce the equations of motion (3.65) and (3.69), as they must. The pair of equations

$$\dot{q}_{\alpha i} = \frac{\partial H}{\partial p_{\alpha i}} \qquad \dot{a}_i = \frac{\delta H}{\delta \pi_i^{\perp}} \tag{3.74}$$

are equivalent to the relations between the canonical momenta and the Lagrangian velocities, whereas the equations

$$\dot{p}_{\alpha i} = -\frac{\partial H}{\partial q_{\alpha i}} \qquad \dot{\pi}_i = -\frac{\delta H}{\delta a_i^{\perp}} \tag{3.75}$$

yield once again the Lorentz force law and the inhomogeneous wave equation for the vector potential with the transverse current density as source.

We express the canonical equations in a form suitable for quantization by introducing a yet more general type of Poisson bracket. If F and G are functionals of the transverse field variables $\boldsymbol{\pi}$ and $\boldsymbol{a}$ and functions of the particle variables p and q, then their Poisson bracket is defined to be

$$\{F, G\} = \iiint \left(\frac{\delta F}{\delta a_i^{\perp}}\frac{\delta G}{\delta \pi_i^{\perp}} - \frac{\delta F}{\delta \pi_i^{\perp}}\frac{\delta G}{\delta a_i^{\perp}}\right) d^3x + \frac{\partial F}{\partial q_i}\frac{\partial G}{\partial p_i} - \frac{\partial F}{\partial p_i}\frac{\partial G}{\partial q_i} \tag{3.76}$$

With this definition the equation of motion for any dynamical variable F (which is a functional of the canonical coordinates and momenta of the field and a function of those of the particles) may be written as

$$\dot{F} = \{F, H\} \tag{3.77}$$

so long as F has no explicit time dependence. In particular, the canonical coordinates and momenta themselves satisfy this equation. If F and G depend on the particle variables only, the definition (3.76) reduces to (3.11), and if they depend on the field variables only, it reduces to (3.52). Thus the fundamental Poisson brackets of the particle variables alone or the transverse field variables alone are as given previously. The more general definition (3.76) implies in addition that the Poisson bracket of any particle variable with any field variable vanishes identically.

3.5 Conservation of energy and momentum; Noether's principle

If the electromagnetic field falls off sufficiently rapidly at infinity for the surface integrals occurring in the energy balance equation (2.53) and the momentum balance equations (2.58) and (2.60) to vanish when the volume V becomes all of space, the system is said to be *closed*. There are then no external agencies that alter the total energy or momentum (although energy and momentum exchanges take place within the system), and we have the three conservation laws:

(i) Conservation of energy, $U+K=\text{constant}$.

(ii) Conservation of linear momentum, $\boldsymbol{G}+\boldsymbol{P}=\text{constant}$.

(iii) Conservation of angular momentum, $\boldsymbol{I}+\boldsymbol{M}=\text{constant}$.

Since the integrals defining U, $\boldsymbol{G}$ and $\boldsymbol{I}$ now extend over all space, the assumption that the charged particles lie within the volume V under consideration is justified. If the criteria for a closed system are to apply even when the field energy and momentum are transformed as in Section 2.6, then a restriction must be placed on the vector and scalar fields $\boldsymbol{v}$ and s that appear in the transformations. For Equation (2.63) implies that the new energy flux through a surface at infinity (or the integral of the divergence of the new Poynting vector over all space) vanishes along with the old only if $\boldsymbol{v}(\boldsymbol{x}, t)$ is such that $\iiint \nabla \cdot \boldsymbol{v}\, \mathrm{d}^3x$ is constant in time. Similarly Equation (2.69) implies that if the net forces and torques associated with both the old and the new Maxwell stress tensors are to vanish at infinity, then $s(\boldsymbol{x}, t)$ must be such that $\iiint \nabla s\, \mathrm{d}^3x$ and $\iiint \nabla\times(s\boldsymbol{x})\, \mathrm{d}^3x$ are constant. It follows then from Equations (2.62) and (2.67) that the transformed quantities U', $\boldsymbol{G}'$ and $\boldsymbol{I}'$ differ from U, $\boldsymbol{G}$ and $\boldsymbol{I}$, respectively, only by constants. The addition of such constants to the total energy and momentum does not affect the content of the conservation laws.

It is of interest to see how the conservation of energy and momentum stems directly, in accordance with Noether's principle (Noether 1918), from the symmetries of the Lagrangian that are continuously connected with the identity. First, since L has no *explicit* dependence on time,

$$\frac{\mathrm{d}L}{\mathrm{d}t}=\iiint\left(\frac{\delta L}{\delta a_i^{\perp}}\,\dot{a}_i+\frac{\delta L}{\delta \dot{a}_i^{\perp}}\,\ddot{a}_i\right)\mathrm{d}^3x+\sum_{\alpha}\left(\frac{\partial L}{\partial q_{\alpha i}}\,\dot{q}_{\alpha i}+\frac{\partial L}{\partial \dot{q}_{\alpha i}}\,\ddot{q}_{\alpha i}\right)$$

$$=\iiint(\dot{\boldsymbol{\pi}}\cdot\dot{\boldsymbol{a}}+\boldsymbol{\pi}\cdot\ddot{\boldsymbol{a}})\,\mathrm{d}^3x+\sum_{\alpha}(\dot{\boldsymbol{p}}_{\alpha}\cdot\dot{\boldsymbol{q}}_{\alpha}+\boldsymbol{p}_{\alpha}\cdot\ddot{\boldsymbol{q}}_{\alpha}) \qquad (3.78)$$

Here the Euler–Lagrange equations and the definitions of the canonical momenta have been used. Since also

$$H=\iiint \boldsymbol{\pi}\cdot\dot{\boldsymbol{a}}\,\mathrm{d}^3x+\sum_\alpha \boldsymbol{p}_\alpha\cdot\dot{\boldsymbol{q}}_\alpha-L \tag{3.79}$$

it follows that

$$\frac{\mathrm{d}H}{\mathrm{d}t}=0 \tag{3.80}$$

so that the Hamiltonian H is a constant of the motion. Now H may be identified with the total energy $U+K$ of the system. If the electric field $\boldsymbol{e}$ is split into its transverse and longitudinal parts, then the total field energy U may be written as the sum of the transverse field energy $U^\perp$, in which $\boldsymbol{e}^\perp$ replaces $\boldsymbol{e}$, and the instantaneous Coulomb energy of the particles. (The infinite self-energy must again be omitted.) Thus the second and third terms in the expression (3.72) for H combine to give the field energy U, while the first term represents the kinetic energy K of the particles. Hence the Hamiltonian is the total energy of the system expressed as a functional or function of the canonical variables. The law of conservation of energy, or the constancy of H, is thus a simple consequence of the fact that L has no explicit dependence on the time (or of the homogeneity of time).

To recover the laws of conservation of momentum using Noether's method, we consider an infinitesimal transformation of the generalized coordinates,

$$\boldsymbol{a}(\boldsymbol{x},t)\to\bar{\boldsymbol{a}}(\boldsymbol{x},t)=\boldsymbol{a}(\boldsymbol{x},t)+\delta\boldsymbol{a}(\boldsymbol{x},t) \tag{3.81}$$

$$q(t)\to\bar{q}(t)=q(t)+\delta q(t) \tag{3.82}$$

and suppose that L is invariant under the transformation, i.e. that

$$L[\bar{\boldsymbol{a}},\dot{\bar{\boldsymbol{a}}};\bar{q},\dot{\bar{q}})\equiv L[\boldsymbol{a},\dot{\boldsymbol{a}};q,\dot{q}) \tag{3.83}$$

This means that the transformation is such that

$$\iiint\left(\frac{\delta L}{\delta a_i^\perp}\delta a_i+\frac{\delta L}{\delta \dot{a}_i^\perp}\frac{\mathrm{d}}{\mathrm{d}t}\delta a_i\right)\mathrm{d}^3x+\sum_\alpha\left(\frac{\partial L}{\partial q_{\alpha i}}\delta q_{\alpha i}+\frac{\partial L}{\partial \dot{q}_{\alpha i}}\frac{\mathrm{d}}{\mathrm{d}t}\delta q_{\alpha i}\right)=0 \tag{3.84}$$

If now the quantity C is defined by

$$C=\iiint \boldsymbol{\pi}\cdot\delta\boldsymbol{a}\,\mathrm{d}^3x+\sum_\alpha \boldsymbol{p}_\alpha\cdot\delta\boldsymbol{q}_\alpha \tag{3.85}$$

then for any natural motion of the system, to which the Euler–Lagrange equations apply,

$$\frac{\mathrm{d}C}{\mathrm{d}t}=0 \tag{3.86}$$

C is thus a constant of the motion associated with the invariant transformation (3.81) and (3.82).

Let us suppose that the transformation is induced by an infinitesimal translation of the reference frame. The new frame $\bar{S}$ is obtained from the old frame S by a rigid displacement (without rotation or reflection) through a distance $\boldsymbol{d}$. We then have

$$\bar{\boldsymbol{x}}=\boldsymbol{x}-\boldsymbol{d} \tag{3.87}$$

where the physical point that has coordinates given by $\boldsymbol{x}=(x_1, x_2, x_3)^{\mathrm{T}}$ in S has coordinates given by $\bar{\boldsymbol{x}}=(\bar{x}_1, \bar{x}_2, \bar{x}_3)^{\mathrm{T}}$ in $\bar{S}$. Here T denotes the transpose, so that $\boldsymbol{x}$ and $\bar{\boldsymbol{x}}$ are column vectors. The vector potential is described by the function $\boldsymbol{a}(\boldsymbol{x}, t)$ in S and by $\bar{\boldsymbol{a}}(\bar{\boldsymbol{x}}, t)$ in $\bar{S}$. Thus†

$$\bar{\boldsymbol{a}}(\bar{\boldsymbol{x}}, t)=\boldsymbol{a}(\boldsymbol{x}, t) \tag{3.88}$$

or, since $\boldsymbol{d}$ is constant and infinitesimal,

$$\begin{aligned}\bar{\boldsymbol{a}}(\boldsymbol{x}, t)&=\boldsymbol{a}(\boldsymbol{x}+\boldsymbol{d}, t)\\&=\boldsymbol{a}(\boldsymbol{x}, t)+\delta\boldsymbol{a}(\boldsymbol{x}, t)\end{aligned} \tag{3.89}$$

where

$$\delta\boldsymbol{a}(\boldsymbol{x}, t)=(\boldsymbol{d}\,.\,\nabla)\boldsymbol{a}(\boldsymbol{x}, t) \tag{3.90}$$

The transformation of the particle coordinates follows from Equation (3.87) and is

$$\bar{\boldsymbol{q}}_\alpha=\boldsymbol{q}_\alpha-\boldsymbol{d} \tag{3.91}$$

so that, for all α,

$$\delta\boldsymbol{q}_\alpha=-\boldsymbol{d} \tag{3.92}$$

Now the Lagrangian L is invariant, in the sense of Equation (3.83), under the combined transformation (3.89) and (3.91). (This is true even for finite displacements $\boldsymbol{d}$.) For since $\boldsymbol{d}$ is constant, the derivatives of the Lagrangian coordinates at a given physical point are unaltered, and the Coulomb energy, since it depends only on the differences between the

† $\boldsymbol{a}(\boldsymbol{x}, t)$ and $\bar{\boldsymbol{a}}(\bar{\boldsymbol{x}}, t)$ refer to the same field, but in the two coordinate systems S and $\bar{S}$ (passive interpretation of the transformation). $\bar{\boldsymbol{a}}(\boldsymbol{x}, t)$ would refer in S to the original field, described in S by $\boldsymbol{a}(\boldsymbol{x}, t)$, displaced through a distance $-\boldsymbol{d}$ (active interpretation).

coordinates of the particles, is also unchanged. Moreover, the integral of the Lagrangian density $\mathscr{L}$, being taken over all space, has the same value for the new as for the old generalized coordinates. Thus C in Equation (3.85) is a constant of the motion when $\delta\boldsymbol{a}$ and $\delta\boldsymbol{q}_\alpha$ are given by Equations (3.90) and (3.92), respectively. An integration by parts then shows that

$$\begin{aligned} C &= -\boldsymbol{d}\cdot\left(\frac{1}{4\pi c}\iiint \boldsymbol{e}^{\perp}\times\boldsymbol{b}\,\mathrm{d}^3x+\sum_\alpha \boldsymbol{p}_\alpha\right) \\ &= -\boldsymbol{d}\cdot\left(\frac{1}{4\pi c}\iiint \boldsymbol{e}\times\boldsymbol{b}\,\mathrm{d}^3x+\sum_\alpha m_\alpha\dot{\boldsymbol{q}}_\alpha\right) \end{aligned} \tag{3.93}$$

and since the direction of the infinitesimal vector $\boldsymbol{d}$ is arbitrary, the conservation of linear momentum is established. We note from Equation (3.93) that the longitudinal momentum of the field combines naturally with the physical momentum of the particles to form the canonical momentum of the particles; the remaining field momentum involves the transverse fields only.

To investigate the conservation of angular momentum, we again adopt the passive point of view and consider a rotation (without reflection or translation) of the coordinate frame. According to Euler's theorem, the orientation of the rotated reference system $\bar{S}$ with respect to the original system S may be specified by a pseudovector $\boldsymbol{n}$ whose direction $\hat{\boldsymbol{n}}$ and magnitude n $(0 \leqslant n \leqslant \pi)$ give the axis and angle of rotation, the sense being that of a right-handed screw if the frames are right-handed and of a left-handed screw if they are left-handed. (Both frames must have the same chirality, since they can be transformed into each other by proper rotations.) The resolutes of a coordinate vector in the two frames are then connected by the equation

$$\bar{x}_i = R_{ij}x_j \tag{3.94}$$

where (Chirgwin and Plumpton 1966)

$$R_{ij} = \delta_{ij}\cos n + \hat{n}_i\hat{n}_j(1-\cos n) + \varepsilon_{ijk}\hat{n}_k \sin n \tag{3.95}$$

The rotation matrix R is orthogonal and has determinant 1. For infinitesimal rotation vectors $\delta\boldsymbol{n}$, Equation (3.94) reduces to

$$\bar{\boldsymbol{x}} = \boldsymbol{x} - \delta\boldsymbol{n}\times\boldsymbol{x} \tag{3.96}$$

The rotation vector $\boldsymbol{n}$ has the same resolutes in $\bar{S}$ and S, as it is an eigenvector of R with eigenvalue 1. We note that $\delta\boldsymbol{n}\times\boldsymbol{x}$ is a true vector, since $\boldsymbol{n}$ is a pseudovector. Equation (3.96), or (3.94), therefore represents the same physical rotation in either a left- or a right-handed frame. The components of the vector potential as described by observers in the two frames $\bar{S}$ and S are related by

$$\bar{a}_i(\bar{\boldsymbol{x}}, t) = R_{ij}a_j(\boldsymbol{x}, t) \tag{3.97}$$

or, since R is orthogonal, by

$$\bar{a}_i(\boldsymbol{x}, t) = R_{ij}a_j(R^{\mathrm{T}}\boldsymbol{x}, t) \tag{3.98}$$

where R^{T} is the transpose of R. For infinitesimal rotations this gives, to first order in δn,

$$\delta \boldsymbol{a}(\boldsymbol{x}, t) = (\delta \boldsymbol{n} \times \boldsymbol{x} \,.\, \nabla)\boldsymbol{a}(\boldsymbol{x}, t) - \delta \boldsymbol{n} \times \boldsymbol{a}(\boldsymbol{x}, t) \tag{3.99}$$

The corresponding equation for the particle coordinates is

$$\delta \boldsymbol{q}_\alpha = -\delta \boldsymbol{n} \times \boldsymbol{q}_\alpha \tag{3.100}$$

The rotational invariance of the Lagrangian may be deduced (for finite as well as infinitesimal rotations) from the orthogonality of the matrix R and from the fact that the Lagrangian density is integrated over all space, so that the region of integration is not affected by the change of variable $\boldsymbol{x} \to R^{\mathrm{T}}\boldsymbol{x}$ in the integrand. We thus obtain another constant of the motion by substituting the expressions (3.99) and (3.100) for $\delta \boldsymbol{a}$ and $\delta \boldsymbol{q}_\alpha$ in Equation (3.85). This gives, after an integration by parts,

$$\begin{aligned} C &= -\delta \boldsymbol{n} \,.\, \left\{ \frac{1}{4\pi c} \iiint \boldsymbol{x} \times (\boldsymbol{e}^{\perp} \times \boldsymbol{b}) \, \mathrm{d}^3 x + \sum_\alpha \boldsymbol{q}_\alpha \times \boldsymbol{p}_\alpha \right\} \\ &= -\delta \boldsymbol{n} \,.\, \left\{ \frac{1}{4\pi c} \iiint \boldsymbol{x} \times (\boldsymbol{e} \times \boldsymbol{b}) \, \mathrm{d}^3 x + \sum_\alpha \boldsymbol{q}_\alpha \times m_\alpha \dot{\boldsymbol{q}}_\alpha \right\} \end{aligned} \tag{3.101}$$

Again the longitudinal field momentum added to the particle physical momentum results in the particle canonical momentum. Since the direction of $\delta \boldsymbol{n}$ is arbitrary, the law of conservation of angular momentum follows from the constancy of C in Equation (3.101). This law is thus a consequence of the invariance of the Lagrangian under proper rotations (or of the isotropy of space), just as the law of conservation of linear momentum is a consequence of its invariance under translations (or of the homogeneity of space).

REFERENCES

Bohr, N. (1923). "Über die Anwendung der Quantentheorie auf den Atombau. I. Die Grundpostulate der Quantentheorie". *Zeitzschrift für Physik* **13,** 117.

Chirgwin, B. H. and Plumpton, C. (1966). "A Course of Mathematics for Engineers and Scientists", Vol. 6. Pergamon, Oxford.

Dirac, P. A. M. (1925). "The fundamental equations of quantum mechanics". *Proc. R. Soc. Lond.* A **109,** 642.

Dirac, P. A. M. (1958). "Generalized Hamiltonian dynamics". *Proc. R. Soc. Lond.* A **246,** 326.

Lanczos, C. (1970). "The Variational Principles of Mechanics". University of Toronto Press.

Noether, E. (1918). "Invariante Variationsprobleme". *Nach Ges. Wiss. Göttingen,* 253.

Volterra, V. (1959). "Theory of Functionals and of Integral and Integro-differential Equations". Dover, New York.

Chapter 4

Canonical Quantization

4.1 Introduction

In quantum mechanics the state of a dynamical system at any instant is specified by a non-zero vector, or, more properly, a non-zero ray, in an abstract Hilbert space, and the dynamical variables for the system are represented by linear operators on this space. A vector corresponding to a state S will be written, in Dirac's notation, as $|S\rangle$ and its Hermitian adjoint, which belongs to the dual space, as $\langle S|$. The ray corresponding to the state S consists of all vectors $c\,|S\rangle$, where c is any non-zero complex number. The Hilbert space† is a linear vector space on which is defined a Hermitian scalar product $\langle R \mid S\rangle$ which is linear in $|S\rangle$ and such that $\langle R \mid S\rangle = \langle S \mid R\rangle^*$. It follows that the scalar product of a vector with itself must be real; it is also assumed to be positive definite, i.e. $\langle S \mid S\rangle \geqslant 0$ with equality if and only if $|S\rangle$ is the zero vector. The norm of $|S\rangle$, denoted by $\|S\|$, is then defined to be the positive square root of $\langle S \mid S\rangle$. It is often convenient to admit into the space vectors that are labelled by continuous parameters and have infinite norms. In this case the linear space is not strictly a Hilbert space, for which the norms must be finite. We shall, however, follow a common usage and call it a Hilbert space. Only those vectors with finite norms correspond to physically realizable states. If such vectors are divided by their norms, the resulting vectors are normalized to unity. The *unit* ray corresponding to a physically realizable state consists of vectors normalized to unity and thus differing from each other by multiplicative phase factors. In the Schrödinger picture of the motion the dynamical variables are constant in time and the state vector changes in accordance with Schrödinger's equation, so long as no measurements are made on the system. In the Heisenberg picture the state vector is constant in time and the dynamical variables change in accordance with Heisenberg's equations. The two pictures are related by a unitary transformation

† For the further postulates of completeness and separability see von Neumann (1955).

generated by the Hamiltonian operator H, which, in the absence of time-dependent external fields, has no explicit dependence on time.

For a system that has a classical analogue, the linear operators corresponding to canonical dynamical variables must satisfy the canonical commutation relations. These are obtained (Dirac 1925) by ascribing to the commutator bracket of two such linear operators a value equal to $i\hbar$ times that of the corresponding classical Poisson bracket. Here $\hbar$ is Planck's constant divided by 2π and the commutator bracket $[A, B]$ of two operators A and B is defined to be $AB-BA$. Since the fundamental Poisson brackets (i.e. those involving the canonical variables among themselves) are constants, the canonical commutator brackets are multiples of the identity operator. It is easy to verify that the commutator brackets have all the algebraic properties of the classical Poisson brackets. Conversely, any "quantum Poisson brackets" possessing these properties must (Dirac 1958) be a constant times the corresponding commutator brackets. If, further, the quantum Poisson bracket of two Hermitian operators is to be Hermitian (by analogy with the reality of the classical Poisson bracket of two real functions), then the constant must be of the form $1/(i\hbar)$, where $\hbar$ is real. The constant $\hbar$ must then have the value stated above, for the theory to agree with experiment. These considerations are also in agreement with Bohr's correspondence principle. The matrix elements of the quantum Poisson brackets are asymptotically equal, in the region of large quantum numbers, to the harmonic components of the corresponding classical Poisson brackets (Dirac 1925, Heisenberg 1930).

PROBLEM

Verify the following relations between the commutator brackets of linear operators.

(i) Skew-symmetry: $[A, B]=-[B, A]$.

(ii) Product rules: $[A_1A_2, B]=[A_1, B]A_2+A_1[A_2, B]$,
$[A, B_1B_2]=[A, B_1]B_2+B_1[A, B_2]$.

(iii) Linearity: $[A_1+A_2, B]=[A_1, B]+[A_2, B]$,
$[A, B_1+B_2]=[A, B_1]+[A, B_2]$,
$[A, cB]=c[A, B]=[cA, B]$,

where c is a number.

(iv) Jacobi's identity:

$$[A,[B, C]]+[B,[C, A]]+[C,[A, B]]=0.$$

4.2 Equations of motion

The canonical dynamical variables for the complete system consisting of the radiation field and a fixed number of non-relativistic charged particles are the coordinates $\boldsymbol{q}_\alpha$ of the particles together with their conjugate momenta $\boldsymbol{p}_\alpha$ and the "coordinates" $\boldsymbol{a}(\boldsymbol{x})$ of the field together with their conjugate momenta $\boldsymbol{\pi}(\boldsymbol{x})$. In accordance with the general quantum mechanical formalism, these are now to be interpreted as operators in Hilbert space. Thus there are six operators $q_{\alpha i}$ and $p_{\alpha i}$ $(i=1, 2, 3)$ associated with each particle α and similarly there are six operators $a_i(\boldsymbol{x})$ and $\pi_i(\boldsymbol{x})$ $(i=1, 2, 3)$ associated with each field point $\boldsymbol{x}$. These are the Schrödinger picture operators, since, as the notation indicates, they have no dependence on time. The corresponding Heisenberg picture operators may be written as $\boldsymbol{q}_\alpha(t)$, $\boldsymbol{p}_\alpha(t)$, $\boldsymbol{a}(\boldsymbol{x}, t)$ and $\boldsymbol{\pi}(\boldsymbol{x}, t)$, respectively. It should be noted that the coordinate vector $\boldsymbol{x}$ is still a c-number parameter labelling the field points, just like the label α for the charged particles or the label t for the time. The particle coordinates $\boldsymbol{q}_\alpha$, on the other hand, become q-numbers. The Schrödinger operators are to satisfy the canonical commutation relations, namely

$$\begin{aligned} &[q_{\alpha i}, q_{\beta j}]=0 \qquad [p_{\alpha i}, p_{\beta j}]=0 \\ &[q_{\alpha i}, p_{\beta j}]=i\hbar\, \delta_{\alpha\beta}\, \delta_{ij} \end{aligned} \tag{4.1}$$

for the particle variables and

$$\begin{aligned} &[a_i(\boldsymbol{x}), a_j(\boldsymbol{x}')]=0 \qquad [\pi_i(\boldsymbol{x}), \pi_j(\boldsymbol{x}')]=0 \\ &[a_i(\boldsymbol{x}), \pi_j(\boldsymbol{x}')]=i\hbar\, \delta^{\perp}_{ij}(\boldsymbol{x}-\boldsymbol{x}') \end{aligned} \tag{4.2}$$

for the field variables. These equations are to hold for all α and β and all i and j. In addition, the commutator bracket of any particle variable with any field variable is to vanish. The last of Equations (4.2) is in agreement with the transverse nature of $\boldsymbol{a}$ and $\boldsymbol{\pi}$, because of the appearance of the transverse delta dyadic on its right-hand side. Thus the Coulomb gauge condition is satisfied from the outset and does not have to be derived from the equations of motion or imposed as a subsidiary constraint.† The canonical commutation relations are also satisfied by the Heisenberg dynamical variables, provided all operators are evaluated at the same time t. This follows from Equations (4.1) and (4.2) and the relation

† This is in contrast to the Gupta–Bleuler formulation of covariant quantum electrodynamics in which the Lorentz gauge condition as an operator identity is incompatible with the commutation relations and appears instead as a restriction on the state vectors.

between the two pictures. If these latter are supposed to coincide at $t=0$, then

$$\Omega(t)=e^{(i/\hbar)Ht}\Omega e^{-(i/\hbar)Ht} \tag{4.3}$$

where H is the Hamiltonian operator and $\Omega(t)$ is the Heisenberg operator corresponding to the Schrödinger operator Ω. Thus $\Omega(t)$ satisfies the Heisenberg equation

$$\dot{\Omega}(t)=\frac{1}{i\hbar}[\Omega(t), H] \tag{4.4}$$

which is the quantum analogue of the classical equation (3.77). The non-equal-time commutation relations for the Heisenberg operators cannot be written down *a priori*, but must be obtained by solving the Heisenberg equations for the system. The Schrödinger picture operator corresponding to $\dot{\Omega}(t)$ is given by

$$\dot{\Omega}=\frac{1}{i\hbar}[\Omega, H] \tag{4.5}$$

In this picture neither Ω nor $\dot{\Omega}$ have any time dependence but $\dot{\Omega}$ is such that

$$\langle S_1(t)|\,\dot{\Omega}\,|S_2(t)\rangle=\frac{\mathrm{d}}{\mathrm{d}t}\langle S_1(t)|\,\Omega\,|S_2(t)\rangle \tag{4.6}$$

where $|S_1(t)\rangle$ and $|S_2(t)\rangle$ are any two solutions of the time-dependent Schrödinger equation

$$i\hbar\frac{\mathrm{d}}{\mathrm{d}t}|S(t)\rangle=H\,|S(t)\rangle \tag{4.7}$$

For the Hamiltonian operator we shall take over the expression (3.72) from the classical theory, namely

$$H=\frac{1}{8\pi}\iiint\{(4\pi c\boldsymbol{\pi})^2+(\nabla\times\boldsymbol{a})^2\}\,\mathrm{d}^3x + \sum_\alpha\frac{1}{2m_\alpha}\left\{\boldsymbol{p}_\alpha-\frac{e_\alpha}{c}\boldsymbol{a}(\boldsymbol{q}_\alpha)\right\}^2+\frac{1}{2}\sum_{\alpha\neq\beta}\frac{e_\alpha e_\beta}{|\boldsymbol{q}_\alpha-\boldsymbol{q}_\beta|} \tag{4.8}$$

and verify that this leads to the expected equations of motion. Since the canonical dynamical variables in the classical theory are real quantities, they are assumed in the quantum theory to be represented by Hermitian operators. This assumption is consistent with the canonical commutation relations. It follows then that H is also a Hermitian operator. We note that there is no ambiguity as to the order of the operators $\boldsymbol{p}_\alpha$ and $\boldsymbol{q}_\alpha$ in Equation (4.8). This is a direct consequence of the Coulomb gauge condition on the vector potential, since

$$[p_{\alpha i}, a_i(\boldsymbol{q}_\alpha)] = -i\hbar \frac{\partial a_i}{\partial q_{\alpha i}} = 0 \tag{4.9}$$

We note also that, since H commutes with the unitary operator $\exp(-iHt/\hbar)$, it is immaterial whether the Schrödinger or the Heisenberg operators are used on the right-hand side of Equation (4.8). Thus H is the same in either picture and is a constant of the motion.

In the derivation of the equations of motion we shall, for economy of notation, use the Schrödinger picture, so that all operators are constant in time and the operator $\dot{\Omega}$ corresponding to Ω is defined by Equation (4.5). As in the classical theory, and as a consequence of the Coulomb gauge, the transverse fields are given in terms of the vector potential by

$$\boldsymbol{e}^{\perp} = -\frac{1}{c}\dot{\boldsymbol{a}} \qquad \boldsymbol{b} = \nabla \times \boldsymbol{a} \tag{4.10}$$

and the longitudinal electric field is a definite function of the particle coordinates:

$$\boldsymbol{e}^{\|}(\boldsymbol{x}) = -\nabla \sum_\alpha \frac{e_\alpha}{|\boldsymbol{x} - \boldsymbol{q}_\alpha|} \tag{4.11}$$

Using the canonical commutation relations and the expression (4.8) for the Hamiltonian, we obtain as the equation of motion for the vector potential evaluated at the field point $\boldsymbol{x}$:

$$\begin{aligned} \dot{a}_i(\boldsymbol{x}) &= 4\pi c^2 \iiint \pi_j(\boldsymbol{x}')\, \delta_{ij}^{\perp}(\boldsymbol{x} - \boldsymbol{x}')\, \mathrm{d}^3x' \\ &= 4\pi c^2 \pi_i(\boldsymbol{x}) \end{aligned} \tag{4.12}$$

where the second line follows from the transversality of $\boldsymbol{\pi}$. Similarly, since $\boldsymbol{a}$ is transverse, the equation for $\boldsymbol{\pi}(\boldsymbol{x})$ is

$$\begin{aligned} \dot{\pi}_i(\boldsymbol{x}) &= -\frac{1}{4\pi}[\nabla \times \{\nabla \times \boldsymbol{a}(\boldsymbol{x})\}]_i \\ &\quad + \sum_\alpha \frac{e_\alpha}{m_\alpha c}\left\{p_{\alpha j} - \frac{e_\alpha}{c} a_j(\boldsymbol{q}_\alpha)\right\} \delta_{ij}^{\perp}(\boldsymbol{x} - \boldsymbol{q}_\alpha) \end{aligned} \tag{4.13}$$

If $\boldsymbol{a}$ is evaluated at a particle point $\boldsymbol{q}_\alpha$, there is an extra term in its time derivative, and then

$$\dot{a}_i(\boldsymbol{q}_\alpha) = 4\pi c^2 \pi_i(\boldsymbol{q}_\alpha) + \frac{1}{2m_\alpha}\left[\frac{\partial a_i}{\partial q_{\alpha j}}\left\{p_{\alpha j} - \frac{e_\alpha}{c}\, a_j(\boldsymbol{q}_\alpha)\right\}\right.$$
$$\left. + \left\{p_{\alpha j} - \frac{e_\alpha}{c}\, a_j(\boldsymbol{q}_\alpha)\right\}\frac{\partial a_i}{\partial q_{\alpha j}}\right] \tag{4.14}$$

The equations for the particle coordinates and momenta are

$$\dot{q}_{\alpha i} = \frac{1}{m_\alpha}\left\{p_{\alpha i} - \frac{e_\alpha}{c}\, a_i(\boldsymbol{q}_\alpha)\right\} \tag{4.15}$$

and

$$\dot{p}_{\alpha i} = \frac{e_\alpha}{m_\alpha c}\left\{p_{\alpha j} - \frac{e_\alpha}{c}\, a_j(\boldsymbol{q}_\alpha)\right\}\frac{\partial a_j}{\partial q_{\alpha i}} + e_\alpha \sum_{\beta \neq \alpha} \frac{e_\beta (q_{\alpha i} - q_{\beta i})}{|\boldsymbol{q}_\alpha - \boldsymbol{q}_\beta|^3} \tag{4.16}$$

respectively. Finally, the time derivative of the longitudinal electric field evaluated at a field point is, from Equation (4.11),

$$\dot{e}_i^{\parallel}(\boldsymbol{x}) = -4\pi \sum_\alpha \frac{e_\alpha}{2m_\alpha}\left[\left\{p_{\alpha j} - \frac{e_\alpha}{c}\, a_j(\boldsymbol{q}_\alpha)\right\}\delta_{ij}^{\parallel}(\boldsymbol{x} - \boldsymbol{q}_\alpha)\right.$$
$$\left. + \delta_{ij}^{\parallel}(\boldsymbol{x} - \boldsymbol{q}_\alpha)\left\{p_{\alpha j} - \frac{e_\alpha}{c}\, a_j(\boldsymbol{q}_\alpha)\right\}\right] \tag{4.17}$$

where the expression (see Appendix C)

$$\delta_{ij}^{\parallel}(\boldsymbol{x}) = -\frac{1}{4\pi}\frac{\partial^2}{\partial x_i\, \partial x_j}\frac{1}{x} \tag{4.18}$$

for the longitudinal delta dyadic has been used.

Because of the relations (4.10) and the Coulomb gauge condition, the source-free Maxwell–Lorentz equations

$$\nabla \cdot \boldsymbol{b} = 0 \qquad \nabla \times \boldsymbol{e}^{\perp} = -\frac{1}{c}\dot{\boldsymbol{b}} \tag{4.19}$$

are identically satisfied. Furthermore, the divergence of the longitudinal electric field is given immediately from Equation (4.11) as

$$\nabla \cdot \boldsymbol{e}^{\parallel}(\boldsymbol{x}) = 4\pi \sum_\alpha e_\alpha\, \delta(\boldsymbol{x} - \boldsymbol{q}_\alpha) \tag{4.20}$$

The expression (4.17) for the time derivative of this field leads, in view of Equation (4.15), to

$$0 = \frac{4\pi}{c}\boldsymbol{j}^{\parallel} + \frac{1}{c}\dot{\boldsymbol{e}}^{\parallel} \tag{4.21}$$

where the longitudinal current density is given by

$$j_i^{\parallel}(\boldsymbol{x}) = \frac{1}{2}\sum_{\alpha} e_{\alpha}\{\dot{q}_{\alpha j}\,\delta_{ij}^{\parallel}(\boldsymbol{x}-\boldsymbol{q}_{\alpha}) + \delta_{ij}^{\parallel}(\boldsymbol{x}-\boldsymbol{q}_{\alpha})\dot{q}_{\alpha j}\} \tag{4.22}$$

The transverse counterpart of Equation (4.21) follows from the equations of motion (4.12), (4.13) and (4.15). Thus

$$\frac{1}{c^2}\ddot{a}_i(\boldsymbol{x}) = -[\nabla\times\{\nabla\times\boldsymbol{a}(\boldsymbol{x})\}]_i + \frac{4\pi}{c}\sum_{\alpha} e_{\alpha}\dot{q}_{\alpha j}\,\delta_{ij}^{\perp}(\boldsymbol{x}-\boldsymbol{q}_{\alpha}) \tag{4.23}$$

which, in conjunction with Equations (4.10), gives

$$\nabla\times\boldsymbol{b} = \frac{4\pi}{c}\boldsymbol{j}^{\perp} + \frac{1}{c}\dot{\boldsymbol{e}}^{\perp} \tag{4.24}$$

where the transverse current density is

$$j_i^{\perp}(\boldsymbol{x}) = \sum_{\alpha} e_{\alpha}\dot{q}_{\alpha j}\,\delta_{ij}^{\perp}(\boldsymbol{x}-\boldsymbol{q}_{\alpha}) \tag{4.25}$$

In summary, the Maxwell–Lorentz field equations are valid as operator equations with the charge and current density operators

$$\rho(\boldsymbol{x}) = \sum_{\alpha} e_{\alpha}\,\delta(\boldsymbol{x}-\boldsymbol{q}_{\alpha}) \tag{4.26}$$

$$\boldsymbol{j}(\boldsymbol{x}) = \frac{1}{2}\sum_{\alpha} e_{\alpha}\{\dot{\boldsymbol{q}}_{\alpha}\,\delta(\boldsymbol{x}-\boldsymbol{q}_{\alpha}) + \delta(\boldsymbol{x}-\boldsymbol{q}_{\alpha})\dot{\boldsymbol{q}}_{\alpha}\} \tag{4.27}$$

The operator equation of motion for a material particle under the influence of the electromagnetic field is obtained from Equations (4.14)–(4.16) and is

$$\begin{aligned} m_{\alpha}\ddot{q}_{\alpha i} &= \frac{e_{\alpha}}{c}\dot{q}_{\alpha j}\frac{\partial a_j}{\partial q_{\alpha i}} + e_{\alpha}e_i^{\parallel}(\boldsymbol{q}_{\alpha}) \\ &\quad - \frac{e_{\alpha}}{2c}\left(\dot{q}_{\alpha j}\frac{\partial a_i}{\partial q_{\alpha j}} + \frac{\partial a_i}{\partial q_{\alpha j}}\dot{q}_{\alpha j}\right) + e_{\alpha}e_i^{\perp}(\boldsymbol{q}_{\alpha}) \end{aligned} \tag{4.28}$$

This may be written as

$$m_{\alpha}\ddot{\boldsymbol{q}}_{\alpha} = e_{\alpha}\left[\boldsymbol{e}(\boldsymbol{q}_{\alpha}) + \frac{1}{2}\left\{\frac{\dot{\boldsymbol{q}}_{\alpha}}{c}\times\boldsymbol{b}(\boldsymbol{q}_{\alpha}) - \boldsymbol{b}(\boldsymbol{q}_{\alpha})\times\frac{\dot{\boldsymbol{q}}_{\alpha}}{c}\right\}\right] \tag{4.29}$$

in which form it is recognizable as Newton's law with the Lorentz force.

The foregoing analysis emphasizes once again that there is no ambiguity in the order of the non-commuting operators $p_{\alpha i}$ and $q_{\alpha i}$. Because of the transverse nature of $\boldsymbol{a}$, this order is immaterial not only in the

Hamiltonian but also, e.g., Equation (4.16), and because of the transverse nature of $\delta_{ij}^{\perp}(\boldsymbol{x})$, it is immaterial also in the expression for the transverse current density. In the expressions for the longitudinal current density and the Lorentz force the order is significant, but is completely prescribed by the formalism.

4.3 Photons

FOURIER ANALYSIS OF THE FIELD

To exhibit the particle properties of the quantized field, it is convenient to consider the Fourier transforms of the field operators, rather than the field operators themselves. We work as before in the Schrödinger picture and define the Fourier transform $\tilde{\boldsymbol{a}}(\boldsymbol{k})$ of $\boldsymbol{a}(\boldsymbol{x})$ by

$$\tilde{\boldsymbol{a}}(\boldsymbol{k}) = \frac{1}{(2\pi)^{3/2}} \iiint \boldsymbol{a}(\boldsymbol{x}) e^{-i\boldsymbol{k}.\boldsymbol{x}} \, \mathrm{d}^3 x \tag{4.30}$$

and assume that the Fourier inversion theorem holds, so that

$$\boldsymbol{a}(\boldsymbol{x}) = \frac{1}{(2\pi)^{3/2}} \iiint \tilde{\boldsymbol{a}}(\boldsymbol{k}) e^{i\boldsymbol{k}.\boldsymbol{x}} \, \mathrm{d}^3 k \tag{4.31}$$

The Fourier transform $\tilde{\boldsymbol{\pi}}(\boldsymbol{k})$ of $\boldsymbol{\pi}(\boldsymbol{x})$ is defined in a similar way. Since $\boldsymbol{a}$ and $\boldsymbol{\pi}$ are Hermitian as well as linear, we have

$$\tilde{\boldsymbol{a}}^{\dagger}(\boldsymbol{k}) = \tilde{\boldsymbol{a}}(-\boldsymbol{k}) \qquad \tilde{\boldsymbol{\pi}}^{\dagger}(\boldsymbol{k}) = \tilde{\boldsymbol{\pi}}(-\boldsymbol{k}) \tag{4.32}$$

where the dagger symbol denotes the Hermitian adjoint. It follows also that since $\boldsymbol{a}$ and $\boldsymbol{\pi}$ are transverse, in the sense that $\nabla \cdot \boldsymbol{a} = 0$ and $\nabla \cdot \boldsymbol{\pi} = 0$, then

$$\boldsymbol{k} \cdot \tilde{\boldsymbol{a}}(\boldsymbol{k}) = 0 \qquad \boldsymbol{k} \cdot \tilde{\boldsymbol{\pi}}(\boldsymbol{k}) = 0 \tag{4.33}$$

i.e. $\tilde{\boldsymbol{a}}$ and $\tilde{\boldsymbol{\pi}}$ have components perpendicular to $\boldsymbol{k}$ only. The commutation relations satisfied by $\tilde{\boldsymbol{a}}$ and $\tilde{\boldsymbol{\pi}}$ may be deduced from the canonical commutation relations satisfied by $\boldsymbol{a}$ and $\boldsymbol{\pi}$ and from the definitions of $\tilde{\boldsymbol{a}}$ and $\tilde{\boldsymbol{\pi}}$. The required relations are

$$[\tilde{a}_i(\boldsymbol{k}), \tilde{a}_j(\boldsymbol{k}')] = 0 \qquad [\tilde{\pi}_i(\boldsymbol{k}), \tilde{\pi}_j(\boldsymbol{k}')] = 0 \tag{4.34}$$

and

$$[\tilde{a}_i(\boldsymbol{k}), \tilde{\pi}_j(\boldsymbol{k}')] = i\hbar(\delta_{ij} - \hat{k}_i \hat{k}_j)\, \delta(\boldsymbol{k} + \boldsymbol{k}') \tag{4.35}$$

In deriving the last relation, the Fourier decomposition (C.6) of the transverse delta dyadic $\delta_{ij}^{\perp}(\boldsymbol{x})$ has also been used. We now introduce a more useful set of operators $\boldsymbol{c}(\boldsymbol{k})$ through the definition

$$\boldsymbol{c}(\boldsymbol{k})=\left(\frac{k}{8\pi\hbar c}\right)^{1/2}\tilde{\boldsymbol{a}}(\boldsymbol{k})+i\left(\frac{2\pi c}{\hbar k}\right)^{1/2}\tilde{\boldsymbol{\pi}}(\boldsymbol{k}) \tag{4.36}$$

where the square root factors are chosen for later convenience. The operators $\boldsymbol{c}(\boldsymbol{k})$, which, like $\tilde{\boldsymbol{a}}$ and $\tilde{\boldsymbol{\pi}}$, are not Hermitian, satisfy the commutation relations

$$[c_i(\boldsymbol{k}),\, c_j(\boldsymbol{k}')]=0=[c_i^{\dagger}(\boldsymbol{k}),\, c_j^{\dagger}(\boldsymbol{k}')] \tag{4.37}$$

and

$$[c_i(\boldsymbol{k}),\, c_j^{\dagger}(\boldsymbol{k}')]=(\delta_{ij}-\hat{k}_i\hat{k}_j)\,\delta(\boldsymbol{k}-\boldsymbol{k}') \tag{4.38}$$

as follows from the relations (4.34) and (4.35) for $\tilde{\boldsymbol{a}}$ and $\tilde{\boldsymbol{\pi}}$. Since $\tilde{\boldsymbol{a}}(\boldsymbol{k})$ and $\tilde{\boldsymbol{\pi}}(\boldsymbol{k})$ are both orthogonal to $\boldsymbol{k}$, the same holds true for $\boldsymbol{c}(\boldsymbol{k})$. Let $\hat{\boldsymbol{e}}^{(1)}(\boldsymbol{k})$ and $\hat{\boldsymbol{e}}^{(2)}(\boldsymbol{k})$ be two unit vectors orthogonal to each other and to $\boldsymbol{k}$ and let them be continuous functions of $\boldsymbol{k}$. Also let $\hat{\boldsymbol{e}}^{(3)}(\boldsymbol{k})=\hat{\boldsymbol{k}}$. The vectors $\hat{\boldsymbol{e}}^{(1)}$ and $\hat{\boldsymbol{e}}^{(2)}$ may be complex and the orthonormality relations (with a Hermitian inner product) are

$$\hat{\boldsymbol{e}}^{(\lambda)*}(\boldsymbol{k})\cdot\hat{\boldsymbol{e}}^{(\lambda')}(\boldsymbol{k})=\delta_{\lambda\lambda'} \qquad (\lambda,\lambda'=1,2,3) \tag{4.39}$$

where the asterisk denotes the complex conjugate. The completeness relations are then

$$\sum_{\lambda=1}^{3}\hat{e}_i^{(\lambda)*}(\boldsymbol{k})\hat{e}_j^{(\lambda)}(\boldsymbol{k})=\delta_{ij} \qquad (i,j=1,2,3) \tag{4.40}$$

or, equivalently,

$$\sum_{\lambda=1}^{2}\hat{e}_i^{(\lambda)*}(\boldsymbol{k})\hat{e}_j^{(\lambda)}(\boldsymbol{k})=\delta_{ij}-\hat{k}_i\hat{k}_j \tag{4.41}$$

which gives the sum over the transverse vectors only. Either of Equations (4.39) and (4.40) states that the nine quantities $\hat{e}_i^{(\lambda)}$ form a 3×3 unitary matrix, which has one real column (if i labels the rows and λ the columns). If $\hat{\boldsymbol{e}}^{(1)}$ and $\hat{\boldsymbol{e}}^{(2)}$ are real, then the matrix is real and orthogonal. Since $\boldsymbol{c}(\boldsymbol{k})$ is perpendicular to $\boldsymbol{k}$, it can be expanded in terms of the transverse unit vectors alone:

$$\boldsymbol{c}(\boldsymbol{k})=\sum_{\lambda=1}^{2}a^{(\lambda)}(\boldsymbol{k})\hat{\boldsymbol{e}}^{(\lambda)}(\boldsymbol{k}) \tag{4.42}$$

with the coefficients being given by

$$a^{(\lambda)}(\boldsymbol{k})=\hat{\boldsymbol{e}}^{(\lambda)*}(\boldsymbol{k})\cdot\boldsymbol{c}(\boldsymbol{k}) \qquad \lambda=1,2 \tag{4.43}$$

The commutation rules for the operators $a^{(\lambda)}(\boldsymbol{k})$ and their Hermitian adjoints $a^{(\lambda)\dagger}(\boldsymbol{k})$ are readily obtained from Equations (4.37), (4.38) and (4.43), and are

$$[a^{(\lambda)}(\boldsymbol{k}), a^{(\lambda')}(\boldsymbol{k}')]=0=[a^{(\lambda)\dagger}(\boldsymbol{k}), a^{(\lambda')\dagger}(\boldsymbol{k}')] \tag{4.44}$$

and

$$[a^{(\lambda)}(\boldsymbol{k}), a^{(\lambda')\dagger}(\boldsymbol{k}')]=\delta_{\lambda\lambda'}\,\delta(\boldsymbol{k}-\boldsymbol{k}') \tag{4.45}$$

These are the so-called Bose commutation relations. As we shall presently see, they enable $a^{(\lambda)}(\boldsymbol{k})$ to be interpreted as an annihilation operator and $a^{(\lambda)\dagger}(\boldsymbol{k})$ as a creation operator for particles (photons) of energy $\hbar ck$, momentum $\hbar\boldsymbol{k}$ and polarization vector $\hat{\boldsymbol{e}}^{(\lambda)}(\boldsymbol{k})$.

The field operators $\boldsymbol{a}(\boldsymbol{x})$ and $\boldsymbol{\pi}(\boldsymbol{x})$ can be expanded directly in terms of the creation and annihilation operators $a^{(\lambda)\dagger}(\boldsymbol{k})$ and $a^{(\lambda)}(\boldsymbol{k})$. Thus

$$\begin{aligned}\boldsymbol{a}(\boldsymbol{x})=\frac{1}{2\pi}\sum_{\lambda=1}^{2}\iiint \mathrm{d}^3k\left(\frac{\hbar c}{k}\right)^{1/2}\{&\hat{\boldsymbol{e}}^{(\lambda)}(\boldsymbol{k})a^{(\lambda)}(\boldsymbol{k})e^{i\boldsymbol{k}.\boldsymbol{x}}\\ &+\hat{\boldsymbol{e}}^{(\lambda)*}(\boldsymbol{k})a^{(\lambda)\dagger}(\boldsymbol{k})e^{-i\boldsymbol{k}.\boldsymbol{x}}\}\end{aligned} \tag{4.46}$$

and

$$\begin{aligned}\boldsymbol{\pi}(\boldsymbol{x})=-\frac{1}{8\pi^2}\sum_{\lambda=1}^{2}\iiint \mathrm{d}^3k\left(\frac{\hbar k}{c}\right)^{1/2} i\{&\hat{\boldsymbol{e}}^{(\lambda)}(\boldsymbol{k})a^{(\lambda)}(\boldsymbol{k})e^{i\boldsymbol{k}.\boldsymbol{x}}\\ &-\hat{\boldsymbol{e}}^{(\lambda)*}(\boldsymbol{k})a^{(\lambda)\dagger}(\boldsymbol{k})e^{-i\boldsymbol{k}.\boldsymbol{x}}\}\end{aligned} \tag{4.47}$$

The corresponding expansions of the transverse electric and magnetic induction fields are then given by

$$\begin{aligned}\boldsymbol{e}^{\perp}(\boldsymbol{x})&=-4\pi c\boldsymbol{\pi}(x)\\ &=\frac{1}{2\pi}\sum_{\lambda=1}^{2}\iiint \mathrm{d}^3k(\hbar ck)^{1/2}i\{\hat{\boldsymbol{e}}^{(\lambda)}(\boldsymbol{k})a^{(\lambda)}(\boldsymbol{k})e^{i\boldsymbol{k}.\boldsymbol{x}}\\ &\quad-\hat{\boldsymbol{e}}^{(\lambda)*}(\boldsymbol{k})a^{(\lambda)\dagger}(\boldsymbol{k})e^{-i\boldsymbol{k}.\boldsymbol{x}}\}\end{aligned} \tag{4.48}$$

and

$$\begin{aligned}\boldsymbol{b}(\boldsymbol{x})&=\nabla\times\boldsymbol{a}(\boldsymbol{x})\\ &=\frac{1}{2\pi}\sum_{\lambda=1}^{2}\iiint \mathrm{d}^3k(\hbar ck)^{1/2}i\{\hat{\boldsymbol{k}}\times\hat{\boldsymbol{e}}^{(\lambda)}(\boldsymbol{k})a^{(\lambda)}(\boldsymbol{k})e^{i\boldsymbol{k}.\boldsymbol{x}}\\ &\quad-\hat{\boldsymbol{k}}\times\hat{\boldsymbol{e}}^{(\lambda)*}(\boldsymbol{k})a^{(\lambda)\dagger}(\boldsymbol{k})e^{-i\boldsymbol{k}.\boldsymbol{x}}\}\end{aligned} \tag{4.49}$$

respectively. Because $\hat{\boldsymbol{e}}^{(\lambda)}(\boldsymbol{k})$ is, for each λ, a continuous function of $\boldsymbol{k}$, the order of the integration over $\boldsymbol{k}$ and the summation over λ in these expressions can be inverted.

IRREDUCIBLE REPRESENTATION OF CREATION AND ANNIHILATION OPERATORS

The commutation relations (4.44) and (4.45) for the operators $a^{(\lambda)}(\boldsymbol{k})$ and $a^{(\lambda)\dagger}(\boldsymbol{k})$ are similar to the commutation relations for the lowering and raising operators that occur in the elementary theory of the harmonic oscillator. The radiation field might thus be considered as an ensemble of harmonic oscillators, one for each mode $\boldsymbol{k}, \lambda$ of the field. Due to the continuous nature of the variable $\boldsymbol{k}$ and the appearance of the Dirac rather than the Kronecker delta on the right-hand side of Equation (4.45), however, we should, using a suitably normalized weight function, integrate the operators $a^{(\lambda)}(\boldsymbol{k})$ over a small region surrounding $\boldsymbol{k}$, in order to obtain exactly the commutation relations for a discrete set of independent oscillators. The ground state of the oscillator for the mode $\boldsymbol{k}, \lambda$ can then be taken to correspond to the absence of any photons for that mode, while the nth excited state ($n = 1, 2, \ldots$), generated by an n-fold application of the appropriate raising or creation operator and having an energy $n\hbar ck$ above the ground state, can be interpreted as corresponding to the presence of n photons, each with energy $\hbar ck$. Now for a finite number of oscillators the Hilbert space is characterized by the requirement that the Hermitian linear operators satisfying the canonical commutation relations form an irreducible set. (A set of linear operators acting on a given space is said to be irreducible if the only linear operators commuting with every operator in the set are multiples of the identity.) For the raising and lowering operators are then determined up to unitary equivalence (von Neumann 1931), i.e. corresponding elements of any two irreducible representations are connected by a unitary transformation. This is not true when an infinite number of degrees of freedom is involved, however, as there are then uncountably many inequivalent irreducible representations of the commutation relations (Wightman and Schweber 1955). Thus in quantum electrodynamics an additional postulate, namely the existence of a vacuum state with certain properties, is needed. This, together with the requirement of irreducibility,† is then sufficient to fix the space.

The vacuum state is assumed to be a unique normalizable state corresponding to a vector $|\text{vac}\rangle$ with the properties

$$\langle \text{vac} \,|\, \text{vac}\rangle = 1 \qquad a^{(\lambda)}(\boldsymbol{k})\,|\text{vac}\rangle = 0 \qquad \text{for all } \boldsymbol{k}, \lambda \tag{4.50}$$

Thus the vacuum is annihilated by all the operators $a^{(\lambda)}(\boldsymbol{k})$. The Hilbert space for the photon system can be built up by using only the properties

† The irreducibility condition is not imposed in all quantum field theories, e.g. in those with so-called superselection rules.

(4.50) and the commutation relations for the creation and annihilation operators. We first define the photon number operator by

$$N = \sum_{\lambda=1}^{2} \iiint d^3k N^{(\lambda)}(\boldsymbol{k}) \tag{4.51}$$

where $N^{(\lambda)}(\boldsymbol{k}) = a^{(\lambda)\dagger}(\boldsymbol{k})a^{(\lambda)}(\boldsymbol{k})$ and is the number operator for mode $\boldsymbol{k}$, λ. Both $N^{(\lambda)}(\boldsymbol{k})$ and N are Hermitian. It is evident that $|\text{vac}\rangle$ is an eigenvector of each $N^{(\lambda)}(\boldsymbol{k})$, and hence of N, with eigenvalue zero—the vacuum is a no-particle state. The one-photon states are obtained by application of the creation operators to the vacuum state. Thus if

$$|\boldsymbol{k}, \lambda\rangle = a^{(\lambda)\dagger}(\boldsymbol{k})\,|\text{vac}\rangle \tag{4.52}$$

then $|\boldsymbol{k}, \lambda\rangle$ is an eigenvector of N with eigenvalue one. This follows from the commutation relations and the definition of the vacuum, since

$$[N^{(\lambda')}(\boldsymbol{k}'), a^{(\lambda)\dagger}(\boldsymbol{k})] = \delta_{\lambda\lambda'}\delta(\boldsymbol{k} - \boldsymbol{k}')a^{(\lambda)\dagger}(\boldsymbol{k}) \tag{4.53}$$

and so

$$\begin{aligned} N\,|\boldsymbol{k}, \lambda\rangle &= \sum_{\lambda'=1}^{2} \iiint d^3k'\, \delta_{\lambda\lambda'}\, \delta(\boldsymbol{k} - \boldsymbol{k}')a^{(\lambda)\dagger}(\boldsymbol{k})\,|\text{vac}\rangle \\ &= |\boldsymbol{k}, \lambda\rangle \end{aligned} \tag{4.54}$$

In a similar way we may deduce the orthonormality properties of the one-photon states. First the effect of an annihilation operator on a one-photon state vector is given by

$$\begin{aligned} a^{(\lambda)}(\boldsymbol{k})\,|\boldsymbol{k}', \lambda'\rangle &= [a^{(\lambda)}(\boldsymbol{k}), a^{(\lambda')\dagger}(\boldsymbol{k}')]\,|\text{vac}\rangle \\ &= \delta_{\lambda\lambda'}\, \delta(\boldsymbol{k} - \boldsymbol{k}')\,|\text{vac}\rangle \end{aligned} \tag{4.55}$$

and hence

$$\begin{aligned} \langle \boldsymbol{k}, \lambda \mid \boldsymbol{k}', \lambda'\rangle &= \langle \text{vac}|\, a^{(\lambda)}(\boldsymbol{k})\,|\boldsymbol{k}', \lambda'\rangle \\ &= \delta_{\lambda\lambda'}\, \delta(\boldsymbol{k} - \boldsymbol{k}') \end{aligned} \tag{4.56}$$

Moreover, since $a^{(\lambda)\dagger}(\boldsymbol{k})$ operating to the left on $\langle\text{vac}|$ gives zero, we have

$$\langle \text{vac} \mid \boldsymbol{k}, \lambda\rangle = \langle \text{vac}|\, a^{(\lambda)\dagger}(\boldsymbol{k})\,|\text{vac}\rangle = 0 \tag{4.57}$$

so that the one-photon subspace, i.e. the space spanned by all the vectors $|\boldsymbol{k}, \lambda\rangle$, is orthogonal to the no-photon subspace, i.e. to the space spanned by $|\text{vac}\rangle$. Any normalized vector $|F\rangle$ in the one-photon subspace has the expansion

$$|F\rangle = \sum_{\lambda=1}^{2} \iiint d^3k\, |\boldsymbol{k}, \lambda\rangle\langle \boldsymbol{k}, \lambda \mid F\rangle \tag{4.58}$$

where the one-photon amplitude $\langle \boldsymbol{k}, \lambda \mid F\rangle$ satisfies

$$\sum_{\lambda=1}^{2} \iiint \mathrm{d}^3k \, |\langle \boldsymbol{k}, \lambda \mid F\rangle|^2 = 1 \tag{4.59}$$

The vectors $|\boldsymbol{k}, \lambda\rangle$ are not normalizable and therefore do not represent physically realizable states. By choosing the amplitude $\langle \boldsymbol{k}, \lambda \mid F\rangle$ to differ from zero only in a small region of $\boldsymbol{k}$-space and for a particular value of λ, however, we can construct physically realizable states that approximate arbitrarily closely to one-photon states with given wave vector and polarization (see Section 6.1).

An n-photon state vector $(n = 1, 2, \ldots)$ is obtained by applying n creation operators in succession to the vacuum state vector:

$$|\boldsymbol{k}_1, \lambda_1; \ldots; \boldsymbol{k}_n, \lambda_n\rangle = a^{(\lambda_1)\dagger}(\boldsymbol{k}_1) \ldots a^{(\lambda_n)\dagger}(\boldsymbol{k}_n) \, |\mathrm{vac}\rangle \tag{4.60}$$

This vector is invariant under any permutation of the indices $(1, \ldots, n)$, since the creation operators all commute. Thus if p is a permutation of $(1, \ldots, n)$, then

$$|\boldsymbol{k}_{p(1)}, \lambda_{p(1)}; \ldots; \boldsymbol{k}_{p(n)}, \lambda_{p(n)}\rangle = |\boldsymbol{k}_1, \lambda_1; \ldots; \boldsymbol{k}_n, \lambda_n\rangle \tag{4.61}$$

where $p(i)$ $(i = 1, \ldots, n)$ is the image of i under p. The properties of the n-photon states may be established by using those of the one-photon states and by induction on n. An annihilation operator acting on an n-photon state vector gives a linear combination of $(n-1)$-photon state vectors. Thus

$$a^{(\lambda)}(\boldsymbol{k}) \, |\boldsymbol{k}_1, \lambda_1; \ldots; \boldsymbol{k}_n, \lambda_n\rangle = \sum_{i=1}^{n} \delta_{\lambda\lambda_i} \delta(\boldsymbol{k} - \boldsymbol{k}_i) \, | \ldots; \boldsymbol{k}_{i-1}, \lambda_{i-1}; \boldsymbol{k}_{i+1}, \lambda_{i+1}; \ldots \rangle \tag{4.62}$$

where the ith $(n-1)$-photon state represented on the right is obtained from the n-photon state represented on the left by removing the ith photon. In the case where $n = 1$, the $(n-1)$- or no-photon state is to be interpreted as the vacuum, in accordance with Equation (4.55). The effect of a creation operator on an n-photon state vector is to produce an $(n+1)$-photon state vector,

$$a^{(\lambda)\dagger}(\boldsymbol{k}) \, |\boldsymbol{k}_1, \lambda_1; \ldots; \boldsymbol{k}_n, \lambda_n\rangle = |\boldsymbol{k}, \lambda; \boldsymbol{k}_1, \lambda_1; \ldots; \boldsymbol{k}_n, \lambda_n\rangle \tag{4.63}$$

as follows directly from the definition (4.60). Equations (4.62) and (4.63) then imply that

$$N \, |\boldsymbol{k}_1, \lambda_1; \ldots; \boldsymbol{k}_n, \lambda_n\rangle = n \, |\boldsymbol{k}_1, \lambda_1; \ldots; \boldsymbol{k}_n, \lambda_n\rangle \tag{4.64}$$

so that any n-photon state is an eigenstate of the number operator with eigenvalue n. The n-photon states have the orthonormality properties

$$\langle \boldsymbol{k}_1, \lambda_1; \ldots; \boldsymbol{k}_n, \lambda_n \mid \boldsymbol{k}'_1, \lambda'_1; \ldots; \boldsymbol{k}'_n, \lambda'_n \rangle = \sum_{p} \delta_{\lambda_1 \lambda'_{p(1)}} \delta(\boldsymbol{k}_1 - \boldsymbol{k}'_{p(1)}) \ldots \delta_{\lambda_n \lambda'_{p(n)}} \delta(\boldsymbol{k}_n - \boldsymbol{k}'_{p(n)}) \quad (4.65)$$

where the sum is over all $n!$ permutations p of $(1, \ldots, n)$. Thus two n-photon states are orthogonal unless they are identical, the order of the indices in the corresponding state vectors being immaterial. Moreover, any n-photon state is orthogonal to the vacuum state and to any m-photon state for which $m \neq n$.

Consider now the vector space spanned by $|\text{vac}\rangle$ and all possible n-photon state vectors for $n = 1, 2, \ldots$. A vector $|F\rangle$ in this space may have a component in the no-photon subspace or in any n-photon subspace:

$$|F\rangle = c_0 |\text{vac}\rangle + \sum_{n=1}^{\infty} \iiint d^3k_1 \ldots \iiint d^3k_n \sum_{\lambda_1=1}^{2} \ldots \sum_{\lambda_n=1}^{2} c(\boldsymbol{k}_1, \lambda_1; \ldots; \boldsymbol{k}_n, \lambda_n) |\boldsymbol{k}_1, \lambda_1; \ldots; \boldsymbol{k}_n, \lambda_n\rangle \quad (4.66)$$

Because of the permutational invariance of the n-photon state vectors, we may choose the coefficients in this expansion so that

$$c(\boldsymbol{k}_{p(1)}, \lambda_{p(1)}; \ldots; \boldsymbol{k}_{p(n)}, \lambda_{p(n)}) = c(\boldsymbol{k}_1, \lambda_1; \ldots; \boldsymbol{k}_n, \lambda_n) \quad (4.67)$$

for any permutation p of $(1, \ldots, n)$. We then obtain, using the orthonormality properties of the basis vectors,

$$\langle \text{vac} \mid F\rangle = c_0 \quad (4.68)$$

and

$$\langle \boldsymbol{k}_1, \lambda_1; \ldots; \boldsymbol{k}_n, \lambda_n \mid F\rangle = \sum_{p} c(\boldsymbol{k}_{p(1)}, \lambda_{p(1)}; \ldots; \boldsymbol{k}_{p(n)}, \lambda_{p(n)}) = n!\, c(\boldsymbol{k}_1, \lambda_1; \ldots; \boldsymbol{k}_n, \lambda_n) \quad (4.69)$$

Thus the expansion for $|F\rangle$ is

$$|F\rangle = |\text{vac}\rangle\langle \text{vac} \mid F\rangle + \sum_{n=1}^{\infty} \frac{1}{n!} \iiint d^3k_1 \ldots \iiint d^3k_n \sum_{\lambda_1=1}^{2} \ldots \sum_{\lambda_n=1}^{2} |\boldsymbol{k}_1, \lambda_1; \ldots; \boldsymbol{k}_n, \lambda_n\rangle\langle \boldsymbol{k}_1, \lambda_1; \ldots; \boldsymbol{k}_n, \lambda_n |F\rangle \quad (4.70)$$

and that for the square of its norm $\|F\|$ is

$$\|F\|^2 = |\langle \text{vac} \mid F\rangle|^2 + \sum_{n=1}^{\infty} \frac{1}{n!} \iiint \mathrm{d}^3 k_1 \ldots \iiint \mathrm{d}^3 k_n \sum_{\lambda_1=1}^{2} \ldots \sum_{\lambda_n=1}^{2} |\langle \boldsymbol{k}_1, \lambda_1; \ldots; \boldsymbol{k}_n, \lambda_n \mid F\rangle|^2 \quad (4.71)$$

F is a physically realizable state if $0 < \|F\|^2 < \infty$. The vector $|F\rangle$ may then be assumed to be normalized to unity, in which case $|\langle \boldsymbol{k}_1, \lambda_1; \ldots; \boldsymbol{k}_n, \lambda_n \mid F\rangle|^2 \, \mathrm{d}^3 k_1 \ldots \mathrm{d}^3 k_n$ is the probability for finding n photons, one with wavevector within $\mathrm{d}^3 k_1$ of $\boldsymbol{k}_1$ and with polarization $\lambda_1, \ldots,$ and one with wavevector within $\mathrm{d}^3 k_n$ of $\boldsymbol{k}_n$ and with polarization λ_n, if the system is known to be in state F before a measurement of the number of photons and of their wavevectors and polarization is made.† Moreover, $|\langle \text{vac} \mid F\rangle|^2$ is the probability that there are no photons at all present. The n-photon amplitude $\langle \boldsymbol{k}_1, \lambda_1; \ldots; \boldsymbol{k}_n, \lambda_n \mid F\rangle$ is unchanged by any permutation of the indices $(1, \ldots, n)$. Thus photons are bosons, i.e. identical particles described by a wavefunction that is symmetric under permutation of the labels of the particles. It is impossible to determine (and thus meaningless to say) which photon is in which state. The permutational symmetry of the n-photon amplitude is responsible for the appearance of the factors $1/n!$ in the expansions (4.70) and (4.71). The n-photon integration in these expansions is carried out over a $3n$-dimensional $\boldsymbol{k}$-space. This space can be divided into $n!$ regions, each of which is uniquely associated with one of the $n!$ permutations of $(1, \ldots, n)$. For example, the region R_p defined by the condition

$$|\boldsymbol{k}_{\mathrm{p}(1)}| > |\boldsymbol{k}_{\mathrm{p}(2)}| > \ldots > |\boldsymbol{k}_{\mathrm{p}(n)}| > 0 \quad (4.72)$$

can be associated with the permutation p. The $n!$ regions so defined are disjoint and simply connected and together with their boundaries, on which two or more of $\boldsymbol{k}_1, \ldots, \boldsymbol{k}_n$ are equal, comprise the entire $3n$-dimensional space. Because of the symmetry of the n-photon amplitude and because the polarizations are summed over, the contribution from each region R_p to the integral over the whole space is the same—each n-photon vector is counted $n!$ times. The integration could thus be restricted to just one of the regions R_p and the factors $1/n!$ omitted. Since each n-photon vector would then be counted only once,

† This assumes that the $3n$-dimensional volume element $\mathrm{d}^3 k_1 \ldots \mathrm{d}^3 k_n$ lies entirely within one of the regions R_p defined in Equation (4.72). See Section 6.1 for the probability when $\boldsymbol{k}_1, \ldots, \boldsymbol{k}_n$ is on the boundary of such a region.

$\langle \boldsymbol{k}_1, \lambda_1; \ldots; \boldsymbol{k}_n, \lambda_n \mid F\rangle$ is the correct amplitude to use in computing the probabilities. In the following, we shall take the integration over the whole $3n$-dimensional space and retain the factors $1/n!$ as well as the normalization (4.65).

The set of all creation and annihilation operators on the space spanned by the vacuum state vector and by all n-photon state vectors for $n = 1, 2, \ldots$ is irreducible. Let L be a linear operator defined everywhere and commuting with all the creation and annihilation operators. Since L commutes with the annihilation operators,

$$a^{(\lambda)}(\boldsymbol{k})L\,|\text{vac}\rangle = La^{(\lambda)}(\boldsymbol{k})\,|\text{vac}\rangle = 0 \tag{4.73}$$

for all $\boldsymbol{k}$, λ. But the vacuum state is, by hypothesis, unique and characterized by being represented by a normalizable ray that is annihilated by all the operators $a^{(\lambda)}(\boldsymbol{k})$. Thus if $L\,|\text{vac}\rangle$ is not zero, it must be proportional to $|\text{vac}\rangle$. In either case we have

$$L\,|\text{vac}\rangle = c\,|\text{vac}\rangle \tag{4.74}$$

for some complex number c. Then since L commutes with all the creation operators and these are linear,

$$\begin{aligned} L\,|\boldsymbol{k}_1, \lambda_1; \ldots; \boldsymbol{k}_n, \lambda_n\rangle &= a^{(\lambda_1)\dagger}(\boldsymbol{k}_1) \ldots a^{(\lambda_n)\dagger}(\boldsymbol{k}_n)L\,|\text{vac}\rangle \\ &= c\,|\boldsymbol{k}_1, \lambda_1; \ldots; \boldsymbol{k}_n, \lambda_n\rangle \end{aligned} \tag{4.75}$$

Finally, from the expansion (4.70) we obtain, since L also is linear

$$L\,|F\rangle = c\,|F\rangle \tag{4.76}$$

for any $|F\rangle$, i.e. L is c times the identity operator. This is just the condition for the set of creation and annihilation operators to be irreducible.

LINEARLY AND CIRCULARLY POLARIZED PHOTONS

The creation operators $a^{(\lambda)\dagger}(\boldsymbol{k})$ depend on the choice of the orthogonal polarization vectors $\hat{\boldsymbol{e}}^{(\lambda)}(\boldsymbol{k})$. Any two orthogonal polarization bases for a given wave vector $\boldsymbol{k}$ are connected by a two-dimensional unitary transformation:

$$\hat{\boldsymbol{e}}^{(\lambda)\prime}(\boldsymbol{k}) = \sum_{\mu=1}^{2} c_{\lambda\mu}(\boldsymbol{k})\hat{\boldsymbol{e}}^{(\mu)}(\boldsymbol{k}) \qquad \lambda = 1, 2 \tag{4.77}$$

where the $\hat{\boldsymbol{e}}^{(\lambda)\prime}(\boldsymbol{k})$ are new polarization vectors and the coefficients $c_{\lambda\mu}(\boldsymbol{k})$ satisfy the relations

$$\sum_{\nu=1}^{2} c^*_{\lambda\nu}(\boldsymbol{k})c_{\mu\nu}(\boldsymbol{k}) = \delta_{\lambda\mu} = \sum_{\nu=1}^{2} c^*_{\nu\lambda}(\boldsymbol{k})c_{\nu\mu}(\boldsymbol{k}) \tag{4.78}$$

The transformation of the creation operators then follows that of the polarization vectors,

$$a^{(\lambda)'\dagger}(\boldsymbol{k}) = \sum_{\mu=1}^{2} c_{\lambda\mu}(\boldsymbol{k}) a^{(\mu)\dagger}(\boldsymbol{k}) \tag{4.79}$$

as may be seen from Equation (4.43). The coefficients $c_{\lambda\mu}(\boldsymbol{k})$ are uniquely determined by the old and new basis vectors, but not by the old and new polarization states, since these latter determine the basis vectors only to within independent multiplicative phase factors.

A polarization vector $\hat{\boldsymbol{e}}^{(\lambda)}(\boldsymbol{k})$ is said to be *essentially real* if it is of the form $\exp\{i\alpha^{(\lambda)}(\boldsymbol{k})\}\hat{\boldsymbol{\varepsilon}}^{(\lambda)}(\boldsymbol{k})$ where $\hat{\boldsymbol{\varepsilon}}^{(\lambda)}(\boldsymbol{k})$ is a real unit vector orthogonal to $\boldsymbol{k}$ and $\alpha^{(\lambda)}(\boldsymbol{k})$ is a real number. If a polarization vector $\hat{\boldsymbol{e}}^{(1)}(\boldsymbol{k})$ is essentially real, then any orthogonal polarization vector $\hat{\boldsymbol{e}}^{(2)}(\boldsymbol{k})$ is also essentially real. For if $\hat{\boldsymbol{e}}^{(1)}(\boldsymbol{k})$ is given, the orthogonality conditions (4.39) determine $\hat{\boldsymbol{e}}^{(2)}(\boldsymbol{k})$ but for an arbitrary multiplicative phase factor. So if $\hat{\boldsymbol{\varepsilon}}^{(2)}(\boldsymbol{k})$ is a real unit vector orthogonal to $\boldsymbol{k}$ and $\hat{\boldsymbol{\varepsilon}}^{(1)}(\boldsymbol{k})$, and is thus uniquely determined apart from sign, then $\hat{\boldsymbol{e}}^{(2)}(\boldsymbol{k})$ is of the form $\exp\{i\alpha^{(2)}(\boldsymbol{k})\}\hat{\boldsymbol{\varepsilon}}^{(2)}(\boldsymbol{k})$ for some real $\alpha^{(2)}(\boldsymbol{k})$. If the polarization vectors $\hat{\boldsymbol{e}}^{(\lambda)}(\boldsymbol{k})$, $\lambda = 1, 2$, are essentially real, the photons associated with the creation operators $a^{(\lambda)\dagger}(\boldsymbol{k})$ are said to be linearly polarized. (Classically, the real or imaginary part of $\hat{\boldsymbol{e}}^{(\lambda)}(\boldsymbol{k})\exp\{i(\boldsymbol{k}\,.\,\boldsymbol{x} - kct)\}$ represents a plane wave linearly polarized along $\hat{\boldsymbol{\varepsilon}}^{(\lambda)}(\boldsymbol{k})$.) Changing the phases $\alpha^{(\lambda)}(\boldsymbol{k})$ does not alter the physical states, although the state vectors get multiplied by phase factors. In particular, changing the sign of a polarization vector does not alter the physical state. Orthogonal linearly polarized photon states thus correspond to any two mutually orthogonal transverse *directions*, without regard to the sense. If both the old and new bases in the transformation (4.77) are essentially real, the unitary matrix connecting them takes the form

$$\begin{bmatrix} c_{11} & c_{12} \\ c_{21} & c_{22} \end{bmatrix} = \begin{bmatrix} e^{i\beta} & 0 \\ 0 & e^{i\gamma} \end{bmatrix} \begin{bmatrix} R_{11} & R_{12} \\ R_{21} & R_{22} \end{bmatrix} \tag{4.80}$$

where $\beta(\boldsymbol{k})$ and $\gamma(\boldsymbol{k})$ are real and $[R_{ij}(\boldsymbol{k})]$ is the real orthogonal matrix relating the two real bases $\hat{\boldsymbol{\varepsilon}}^{(1)}(\boldsymbol{k})$, $\hat{\boldsymbol{\varepsilon}}^{(2)}(\boldsymbol{k})$ and $\hat{\boldsymbol{\varepsilon}}^{(1)\prime}(\boldsymbol{k})$, $\hat{\boldsymbol{\varepsilon}}^{(2)\prime}(\boldsymbol{k})$. By choosing the phases β and γ suitably, R can always be made to represent a proper rotation about the axis $\boldsymbol{k}$.

For the description of circularly polarized photons we introduce two polarization vectors $\hat{\boldsymbol{e}}^{(L)}(\boldsymbol{k})$ and $\hat{\boldsymbol{e}}^{(R)}(\boldsymbol{k})$ with the properties

$$\hat{\boldsymbol{k}}\,.\,\hat{\boldsymbol{e}}^{(L/R)}(\boldsymbol{k}) = 0 \qquad \hat{\boldsymbol{e}}^{(L/R)*}(\boldsymbol{k})\,.\,\hat{\boldsymbol{e}}^{(L/R)}(\boldsymbol{k}) = 1 \tag{4.81}$$

and

$$i\hat{\boldsymbol{k}} \times \hat{\boldsymbol{e}}^{(L)}(\boldsymbol{k}) = \hat{\boldsymbol{e}}^{(L)}(\boldsymbol{k}) \qquad i\hat{\boldsymbol{k}} \times \hat{\boldsymbol{e}}^{(R)}(\boldsymbol{k}) = -\hat{\boldsymbol{e}}^{(R)}(\boldsymbol{k}) \tag{4.82}$$

These equations imply that $\hat{\boldsymbol{e}}^{(\mathrm{L})}$ and $\hat{\boldsymbol{e}}^{(\mathrm{R})}$ are orthogonal. The same equations are satisfied by $\exp(i\alpha^{(\mathrm{L})})\hat{\boldsymbol{e}}^{(\mathrm{L})}$ and $\exp(i\alpha^{(\mathrm{R})})\hat{\boldsymbol{e}}^{(\mathrm{R})}$ for any real $\alpha^{(\mathrm{L})}$ and $\alpha^{(\mathrm{R})}$, and thus $\hat{\boldsymbol{e}}^{(\mathrm{L})}$ and $\hat{\boldsymbol{e}}^{(\mathrm{R})}$ are not uniquely defined. We shall show, however, that they are defined to within multiplicative phase factors, so that the physical states associated with the corresponding creation operators are uniquely determined. It should be noted that $\hat{\boldsymbol{e}}^{(\mathrm{L})}$ and $\hat{\boldsymbol{e}}^{(\mathrm{R})}$ transform under improper rotations neither as true vectors nor as pseudovectors, since otherwise the two sides of each of Equations (4.82) would have differing transformation properties. Let us choose $\hat{\boldsymbol{\varepsilon}}^{(1)}(\boldsymbol{k})$ and $\hat{\boldsymbol{\varepsilon}}^{(2)}(\boldsymbol{k})$ so that they and $\hat{\boldsymbol{k}}$, or $\hat{\boldsymbol{\varepsilon}}^{(3)}(\boldsymbol{k})$, together form a real orthonormal triad with the same chirality as the reference frame, i.e. so that $[\hat{\varepsilon}_i^{(\lambda)}]$ is a real 3×3 orthogonal matrix with determinant $+1$. Then $\hat{\boldsymbol{k}}\times\hat{\boldsymbol{\varepsilon}}^{(1)}=\hat{\boldsymbol{\varepsilon}}^{(2)}$ and $\hat{\boldsymbol{k}}\times\hat{\boldsymbol{\varepsilon}}^{(2)}=-\hat{\boldsymbol{\varepsilon}}^{(1)}$, irrespectively of the handedness of the reference frame. Since $\boldsymbol{k}\cdot\hat{\boldsymbol{e}}^{(\mathrm{L})}=0$, we have $\hat{\boldsymbol{e}}^{(\mathrm{L})}=c_1\hat{\boldsymbol{\varepsilon}}^{(1)}+c_2\hat{\boldsymbol{\varepsilon}}^{(2)}$ and since $\hat{\boldsymbol{e}}^{(\mathrm{L})*}\cdot\hat{\boldsymbol{e}}^{(\mathrm{L})}=1$, we have $|c_1|^2+|c_2|^2=1$. Then $i\hat{\boldsymbol{k}}\times\hat{\boldsymbol{e}}^{(\mathrm{L})}=\hat{\boldsymbol{e}}^{(\mathrm{L})}$ gives $c_2=ic_1$ and so $|c_1|^2=\frac{1}{2}$. Thus

$$\hat{\boldsymbol{e}}^{(\mathrm{L})}(\boldsymbol{k})=e^{i\alpha^{(\mathrm{L})}}(\boldsymbol{k})\frac{1}{\sqrt{2}}[\hat{\boldsymbol{\varepsilon}}^{(1)}(\boldsymbol{k})+i\hat{\boldsymbol{\varepsilon}}^{(2)}(\boldsymbol{k})] \tag{4.83}$$

where $\alpha^{(\mathrm{L})}(\boldsymbol{k})$ is real, and similarly

$$\hat{\boldsymbol{e}}^{(\mathrm{R})}(\boldsymbol{k})=e^{i\alpha^{(\mathrm{R})}}(\boldsymbol{k})\frac{1}{\sqrt{2}}[\hat{\boldsymbol{\varepsilon}}^{(1)}(\boldsymbol{k})-i\hat{\boldsymbol{\varepsilon}}^{(2)}(\boldsymbol{k})] \tag{4.84}$$

where $\alpha^{(\mathrm{R})}(\boldsymbol{k})$ is real. Hence for a given choice of the basis vectors $\hat{\boldsymbol{\varepsilon}}^{(1)}$ and $\hat{\boldsymbol{\varepsilon}}^{(2)}$, the vectors $\hat{\boldsymbol{e}}^{(\mathrm{L})}$ and $\hat{\boldsymbol{e}}^{(\mathrm{R})}$ are defined to within phase factors by Equations (4.81) and (4.82). Moreover, rotating the two-dimensional basis $\hat{\boldsymbol{\varepsilon}}^{(1)}, \hat{\boldsymbol{\varepsilon}}^{(2)}$ about $\boldsymbol{k}$ by a *proper* rotation (so that the chirality of the triad $\hat{\boldsymbol{\varepsilon}}^{(\lambda)}$, $\lambda=1, 2, 3$, is preserved) has the effect of merely multiplying $\hat{\boldsymbol{e}}^{(\mathrm{L})}$ and $\hat{\boldsymbol{e}}^{(\mathrm{R})}$ by phase factors. If this rotation is through an angle θ and has the same sense as the reference frame (i.e. has the sense of a right-handed screw in a right-handed frame and of a left-handed screw in a left-handed frame, the axis in each case being along $\boldsymbol{k}$), then $\hat{\boldsymbol{e}}^{(\mathrm{L})}$ is multiplied by $e^{-i\theta}$ and $\hat{\boldsymbol{e}}^{(\mathrm{R})}$ by $e^{i\theta}$. The photons associated with the creation operators $a^{(\mathrm{L})\dagger}(\boldsymbol{k})$ or $a^{(\mathrm{R})\dagger}(\boldsymbol{k})$, respectively, are said to be left or right circularly polarized. In the classical description, the electric or magnetic induction field vector carries $\hat{\boldsymbol{\varepsilon}}^{(1)}$ into $\hat{\boldsymbol{\varepsilon}}^{(2)}$ through an angle of $\pi/2$, if the light is left circularly polarized, and carries $\hat{\boldsymbol{\varepsilon}}^{(2)}$ into $\hat{\boldsymbol{\varepsilon}}^{(1)}$ through an angle of $\pi/2$, if the light is right circularly polarized. This statement is independent of the chirality of the reference frame (see Fig. 6). Here the handedness of the light has been made to depend on that of the reference frame, so that light that is called left (right) circularly polarized in one frame is called right (left) circularly polarized in a frame of the opposite handedness.

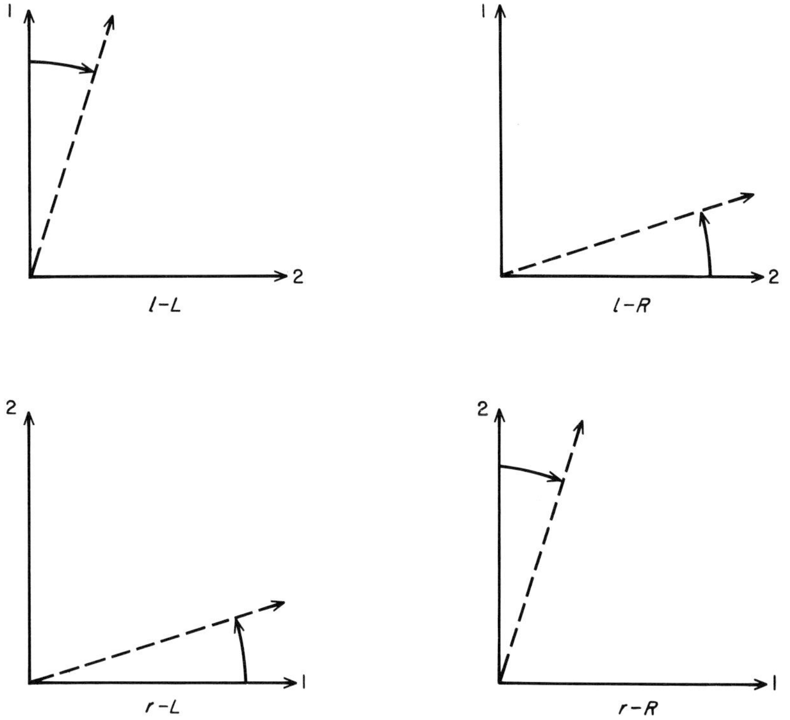

Fig. 6. Illustrating the sense of rotation for left or right circularly polarized light (L or R) in left- or right-handed reference frames (*l* or *r*). The propagation vector points out of the paper.

ZERO POINT ENERGY AND NORMAL ORDERING

Further properties of the photon states may be deduced by considering the operators

$$H_{\mathrm{rad}}=\frac{1}{8\pi}\iiint(\boldsymbol{e}^{\perp 2}+\boldsymbol{b}^2)\,\mathrm{d}^3x \tag{4.85}$$

$$\boldsymbol{P}_{\mathrm{rad}}=\frac{1}{4\pi c}\iiint(\boldsymbol{e}^{\perp}\times\boldsymbol{b})\,\mathrm{d}^3x \tag{4.86}$$

and

$$\boldsymbol{J}_{\mathrm{rad}}=\frac{1}{4\pi c}\iiint\boldsymbol{x}\times(\boldsymbol{e}^{\perp}\times\boldsymbol{b})\,\mathrm{d}^3x \tag{4.87}$$

These represent the part of the total energy, linear momentum and angular momentum, respectively, contained in the transverse field. The commutation relations for these operators follow from those for the components of $\boldsymbol{e}^{\perp}$ and $\boldsymbol{b}$, which in turn follow from the canonical commutation relations (4.2). We have

$$[e_i^{\perp}(\boldsymbol{x}), b_j(\boldsymbol{x}')] = 4\pi i\hbar c\varepsilon_{ijk}\frac{\partial}{\partial x_k'}\delta(\boldsymbol{x}-\boldsymbol{x}') \tag{4.88}$$

with the components of $\boldsymbol{e}^{\perp}$ commuting among themselves, and similarly for the components of $\boldsymbol{b}$. From this we obtain, after integrating by parts and using the divergence condition on $\boldsymbol{e}^{\perp}$ and $\boldsymbol{b}$,

$$\begin{aligned}[H_{\text{rad}}, P_{\text{rad}\,i}] &= -i\hbar\frac{1}{8\pi}\iiint\frac{\partial}{\partial x_i}(\boldsymbol{e}^{\perp 2}+\boldsymbol{b}^2)\,\mathrm{d}^3x \\ &= 0\end{aligned} \tag{4.89}$$

and

$$\begin{aligned}[P_{\text{rad}\,i}, P_{\text{rad}\,j}] &= -i\hbar\frac{1}{4\pi c}\iiint\frac{\partial}{\partial x_j}(\boldsymbol{e}^{\perp}\times\boldsymbol{b})_i\,\mathrm{d}^3x \\ &= 0\end{aligned} \tag{4.90}$$

Because of the vanishing of the commutators, it is possible to measure simultaneously the transverse field energy and all three components of the linear momentum. It is also possible to measure simultaneously the transverse field energy and any component of the angular momentum, since

$$\begin{aligned}[H_{\text{rad}}, J_{\text{rad}\,i}] &= i\hbar\frac{1}{4\pi}\iiint\boldsymbol{x}\times\{\boldsymbol{e}^{\perp}\times(\nabla\times\boldsymbol{e}^{\perp})+\boldsymbol{b}\times(\nabla\times\boldsymbol{b})\}\,|_i\,\mathrm{d}^3x \\ &= i\hbar\varepsilon_{ijk}\oint\!\!\oint_{S_\infty} x_j m_{kl}^{\perp}\,\mathrm{d}S_l \\ &= 0\end{aligned} \tag{4.91}$$

Here S_∞ is a surface bounding all of space and $m^{\perp}$ is the transverse Maxwell stress tensor, i.e. it is the stress tensor (2.54) with $\boldsymbol{e}$ replaced by $\boldsymbol{e}^{\perp}$. The components of $\boldsymbol{J}_{\text{rad}}$ obey the usual angular momentum commutation rules, so that no two of them are simultaneously measurable. Thus

$$[J_{\text{rad}\,i}, (\boldsymbol{e}^{\perp}\times\boldsymbol{b})_j] = i\hbar\{(\boldsymbol{x}\times\nabla)_i(\boldsymbol{e}^{\perp}\times\boldsymbol{b})_j+\varepsilon_{ijk}(\boldsymbol{e}^{\perp}\times\boldsymbol{b})_k\} \tag{4.92}$$

which gives, on dropping a surface term at infinity,

$$[J_{\text{rad}\,i}, J_{\text{rad}\,j}] = i\hbar\varepsilon_{ijk}J_{\text{rad}\,k} \tag{4.93}$$

The commutation relations between the components of $\boldsymbol{J}_{\text{rad}}$ and those of $\boldsymbol{P}_{\text{rad}}$, namely

$$[J_{\text{rad}\,i}, P_{\text{rad}\,j}] = i\hbar\varepsilon_{ijk}P_{\text{rad}\,k} \tag{4.94}$$

may also be obtained directly from Equation (4.92).

The operators H_{rad} and $\boldsymbol{P}_{\text{rad}}$ are diagonal in the representation in which all the number operators $N^{(\lambda)}(\boldsymbol{k})$ are diagonal, i.e. in the plane wave representation. The expansions of H_{rad} and $\boldsymbol{P}_{\text{rad}}$ in terms of the creation and annihilation operators follow from those for $\boldsymbol{e}^{\perp}$ and $\boldsymbol{b}$ and from the relations

$$\{\hat{\boldsymbol{k}}\times\hat{\boldsymbol{e}}^{(\lambda)}(\boldsymbol{k})\}\,.\,\{\hat{\boldsymbol{k}}\times\hat{\boldsymbol{e}}^{(\lambda')*}(\boldsymbol{k})\} = \delta_{\lambda\lambda'} \tag{4.95}$$

$$\{\hat{\boldsymbol{k}}\times\hat{\boldsymbol{e}}^{(\lambda)}(\boldsymbol{k})\}\,.\,\{\hat{\boldsymbol{k}}\times\hat{\boldsymbol{e}}^{(\lambda')}(-\boldsymbol{k})\} = \hat{\boldsymbol{e}}^{(\lambda)}(\boldsymbol{k})\,.\,\hat{\boldsymbol{e}}^{(\lambda')}(-\boldsymbol{k}) \tag{4.96}$$

which may be derived from the orthogonality properties of the polarization vectors. We have

$$\begin{aligned} H_{\text{rad}} &= \frac{1}{2}\sum_{\lambda=1}^{2}\iiint \mathrm{d}^3k\,\hbar ck\{a^{(\lambda)\dagger}(\boldsymbol{k})a^{(\lambda)}(\boldsymbol{k}) + a^{(\lambda)}(\boldsymbol{k})a^{(\lambda)\dagger}(\boldsymbol{k})\} \\ &= \sum_{\lambda=1}^{2}\iiint \mathrm{d}^3k\,\hbar ckN^{(\lambda)}(\boldsymbol{k}) + \text{“constant”} \end{aligned} \tag{4.97}$$

and

$$\begin{aligned} \boldsymbol{P}_{\text{rad}} &= \frac{1}{2}\sum_{\lambda=1}^{2}\iiint \mathrm{d}^3k\,\hbar\boldsymbol{k}\{a^{(\lambda)\dagger}(\boldsymbol{k})a^{(\lambda)}(\boldsymbol{k}) + a^{(\lambda)}(\boldsymbol{k})a^{(\lambda)\dagger}(\boldsymbol{k})\} \\ &= \sum_{\lambda=1}^{2}\iiint \mathrm{d}^3k\,\hbar\boldsymbol{k}N^{(\lambda)}(\boldsymbol{k}) + \text{“constant”} \end{aligned} \tag{4.98}$$

where the commutation relations (4.45) have been used. Unfortunately, however, the “constants” appearing in Equations (4.97) and (4.98) are not well defined, as they are products of the three-dimensional delta function evaluated at the origin and the integral of $\hbar ck$ or $\hbar\boldsymbol{k}$ over all of $\boldsymbol{k}$-space. Although the latter integral might be interpreted as zero when its principal value is taken, the former is divergent, unless a high frequency cut-off is imposed, and gives rise to an infinite “zero-point energy” of the quantized field. This difficulty can be obviated by simply omitting the “constant” terms in Equations (4.97) and (4.98). This procedure amounts to choosing the zero of the energy scale for any radiation oscillator to be that of its ground state, and may be formally implemented as follows. The *normal product* operator **:** **:** embracing a product of creation and annihilation operators is defined to rearrange the product so that all creation operators appear to the left of all annihilation operators. (Because of the commutation rules, the order of creation operators among themselves or of annihilation operators among themselves is immaterial.) In addition, **:** **:** acting on a linear combination of products of creation and annihilation operators is assumed to give the same linear combination of normally ordered products. Thus, e.g.,

$$:aa^{\dagger}: = a^{\dagger}a \qquad :aa^{\dagger} + aa^{\dagger}a: = a^{\dagger}a + a^{\dagger}aa \tag{4.99}$$

where different as may refer to different modes of the field. It follows from the definition that the normally ordered form of a linear combination of products of creation and annihilation operators with real coefficients is Hermitian. Moreover, the vacuum expectation value of a normally ordered product is zero, since

$$a^{(\lambda)}(\boldsymbol{k})\,|\text{vac}\rangle = 0 \qquad \langle\text{vac}|\,a^{(\lambda)\dagger}(\boldsymbol{k}) = 0 \tag{4.100}$$

We now redefine H_{rad} and P_{rad} by putting

$$\begin{aligned} H_{\text{rad}} &= \frac{1}{8\pi}\iiint :(\boldsymbol{e}^{\perp 2} + \boldsymbol{b}^2): \mathrm{d}^3x \\ &= \sum_{\lambda=1}^{2}\iiint \mathrm{d}^3k\,\hbar ck N^{(\lambda)}(\boldsymbol{k}) \end{aligned} \tag{4.101}$$

and

$$\begin{aligned} \boldsymbol{P}_{\text{rad}} &= \frac{1}{4\pi c}\iiint :\boldsymbol{e}^{\perp}\times\boldsymbol{b}: \mathrm{d}^3x \\ &= \sum_{\lambda=1}^{2}\iiint \mathrm{d}^3k\,\hbar\boldsymbol{k} N^{(\lambda)}(\boldsymbol{k}) \end{aligned} \tag{4.102}$$

so that these are Hermitian operators with finite eigenvalues and with lowest eigenvalue zero. Thus

$$H_{\text{rad}}\,|\text{vac}\rangle = 0 \qquad \boldsymbol{P}_{\text{rad}}\,|\text{vac}\rangle = 0 \tag{4.103}$$

and

$$H_{\text{rad}}\,|\boldsymbol{k}_1, \lambda_1; \ldots; \boldsymbol{k}_n, \lambda_n\rangle = \sum_{i=1}^{n} \hbar c\,|\boldsymbol{k}_i|\,|\boldsymbol{k}_1, \lambda_1; \ldots; \boldsymbol{k}_n, \lambda_n\rangle \tag{4.104}$$

$$\boldsymbol{P}_{\text{rad}}\,|\boldsymbol{k}_1, \lambda_1; \ldots; \boldsymbol{k}_n, \lambda_n\rangle = \sum_{i=1}^{n} \hbar\boldsymbol{k}_i\,|\boldsymbol{k}_1, \lambda_1; \ldots; \boldsymbol{k}_n, \lambda_n\rangle \tag{4.105}$$

Because of these relations, a photon with wave vector $\boldsymbol{k}$ is considered as a particle with energy $\hbar ck$ and momentum $\hbar\boldsymbol{k}$, in agreement with the hypotheses of Planck and Einstein. We also redefine the angular momentum operator $\boldsymbol{J}_{\text{rad}}$ by putting

$$\boldsymbol{J}_{\text{rad}} = \frac{1}{4\pi c} \iiint : \boldsymbol{x} \times (\boldsymbol{e}^{\perp} \times \boldsymbol{b}) : \mathrm{d}^3 x \tag{4.106}$$

so that $\boldsymbol{J}_{\text{rad}}$ is also a Hermitian operator that annihilates the vacuum. The components of $\boldsymbol{J}_{\text{rad}}$ are not diagonal in the plane wave representation, since they do not commute with all the components of $\boldsymbol{P}_{\text{rad}}$, although they do commute with H_{rad}. By expanding the field into spherical instead of plane eigenwaves, however, it is possible to find a representation in which H_{rad}, $\boldsymbol{J}^2_{\text{rad}}$ and any one component of $\boldsymbol{J}_{\text{rad}}$ are all diagonal (Heitler 1954). The associated photons have definite values for energy and angular momentum, but not for linear momentum. It should be noted that the redefined operators H_{rad}, $\boldsymbol{P}_{\text{rad}}$ and $\boldsymbol{J}_{\text{rad}}$ obey the same commutation rules as the original operators. This is because they are quadratic functions of the creation and annihilation operators. Roughly speaking, the "constants" by which the original and redefined operators differ are infinite but nevertheless c-numbers. For the same reason, the equations of motion derived in Section 4.2 remain valid.

4.4 Product space for the coupled systems

We have considered the states of the radiation field as a system by itself. These states are represented by non-zero rays in the space generated by the vacuum state vector and by all n-photon state vectors. Let this space be denoted by F. The operators $a^{(\lambda)}(\boldsymbol{k})$ and $a^{(\lambda)\dagger}(\boldsymbol{k})$ for all $\boldsymbol{k}$, λ form an irreducible set of operators on F and satisfy the commutation relations (4.44) and (4.45). Equivalently, the operators $\boldsymbol{a}(\boldsymbol{x})$ and $\boldsymbol{\pi}(\boldsymbol{x})$ for all $\boldsymbol{x}$ form an irreducible set (since any operator that commutes with all of these must also commute with all the creation and annihilation operators) and satisfy the canonical commutation relations (4.2). This, together with the existence of a vacuum state, characterizes F. We can also consider the charged particles as forming a system by itself, as in ordinary quantum mechanics. The states of this system are specified by non-zero rays in a linear space P spanned by, e.g., the eigenvectors $|m\rangle$ of the particle Hamiltonian, given by

$$H_{\text{par}} = \sum_{\alpha} \frac{1}{2m_\alpha} \boldsymbol{p}_\alpha^2 + \frac{1}{2} \sum_{\alpha \neq \beta} \frac{e_\alpha e_\beta}{|\boldsymbol{q}_\alpha - \boldsymbol{q}_\beta|} \tag{4.107}$$

The complete system consists of the interacting field and particles. The interaction part of the complete Hamiltonian (4.8), namely

$$H_{\text{int}} = -\sum_{\alpha} \frac{e_\alpha}{m_\alpha c} \boldsymbol{p}_\alpha \cdot \boldsymbol{a}(\boldsymbol{q}_\alpha) + \sum_{\alpha} \frac{e_\alpha^2}{2m_\alpha c^2} \boldsymbol{a}^2(\boldsymbol{q}_\alpha) \tag{4.108}$$

contains some dynamical variables that refer to the field and some that refer to the particles. For the complete system we use the tensor product space (Dirac 1958) $F \otimes P$, or S, which is generated by the kets $|F; P\rangle \equiv |F\rangle\,|P\rangle$ with $|F\rangle$ in F and $|P\rangle$ in P, a linear combination of such kets being defined by

$$c_1\,|F_1; P\rangle + c_2\,|F_2; P\rangle = \{c_1\,|F_1\rangle + c_2\,|F_2\rangle\}\,|P\rangle \tag{4.109}$$

and by

$$c_1\,|F; P_1\rangle + c_2\,|F; P_2\rangle = |F\rangle\{c_1\,|P_1\rangle + c_2\,|P_2\rangle\} \tag{4.110}$$

so that the distributive law holds. For any linear operator α on F, we define a linear operator (denoted by the same symbol α) on S through the equation

$$\alpha\,|F; P\rangle = \{\alpha\,|F\rangle\}\,|P\rangle \tag{4.111}$$

This defines α everywhere in S, since S is generated by the product kets $|F; P\rangle$ and α is linear. Similarly, if β is a linear operator on P, we define

$$\beta\,|F; P\rangle = |F\rangle\,\{\beta\,|P\rangle\} \tag{4.112}$$

The commutator of two such operators α and β on S must vanish, since

$$(\alpha\beta)\,|F; P\rangle = \{\alpha\,|F\rangle\}\{\beta\,|P\rangle\} = (\beta\alpha)\,|F; P\rangle \tag{4.113}$$

and α and β are linear. It follows that the commutator of any particle variable with any field variable vanishes, as required. The other canonical commutation relations, Equations (4.1) and (4.2), remain unchanged when the operators $\boldsymbol{q}_\alpha$, $\boldsymbol{p}_\alpha$, $\boldsymbol{a}$ and $\boldsymbol{\pi}$ are extended to become operators on the product space. A basis for S is obtained by choosing bases for F and P and taking all the product kets $|F; P\rangle$ where $|F\rangle$ is in the basis for F and $|P\rangle$ is in that for P. If we choose the vacuum state vector and all n-photon state vectors in the plane wave representation as a basis for F and the eigenvectors of the particle Hamiltonian as a basis for P, then the kets $|\text{vac}; m\rangle$ and $|\boldsymbol{k}_1, \lambda_1; \ldots; \boldsymbol{k}_n, \lambda_n; m\rangle$ form a basis for S. Thus if $|S\rangle$ (which is not necessarily of the product form $|F; P\rangle$) represents any physically realizable state of the complete system of field and particles, the probability amplitudes for this state relative to the chosen basis are $\langle m; \text{vac} \mid S\rangle$ and $\langle m; \boldsymbol{k}_1, \lambda_1; \ldots; \boldsymbol{k}_n, \lambda_n \mid S\rangle$, provided $|S\rangle$ is normalized to unity.

It is sometimes convenient to use Schrödinger's representation (Dirac 1958) for the space P that contains the state vectors of the system of charged particles. The position operators $q_{\alpha i}$ on this space are assumed to form a complete set of commuting observables (with eigenvalues ranging continuously from $-\infty$ to ∞), so that their simultaneous eigenkets $|q'\rangle$ form a basis for P. We have

$$\begin{aligned}\langle q' \mid q''\rangle &\equiv \langle \boldsymbol{q}_1'; \ldots ; \boldsymbol{q}_M' \mid \boldsymbol{q}_1''; \ldots ; \boldsymbol{q}_M''\rangle \\ &= \delta(\boldsymbol{q}_1' - \boldsymbol{q}_1'') \ldots \delta(\boldsymbol{q}_M' - \boldsymbol{q}_M'') \\ &\equiv \delta(q' - q'') \end{aligned} \tag{4.114}$$

and

$$q_{\alpha i}\,|q'\rangle = q'_{\alpha i} \mid q'\rangle \tag{4.115}$$

if M is the number of particles. The requirement that the $q_{\alpha i}$ be diagonal does not fix the representation completely, since the basis vectors $|q'\rangle$ can be multiplied by phase factors that vary with the eigenvalues q'. The *standard ket* for a representation is denoted by $\rangle$. This has representative unity for every basis vector, i.e. $\langle q'|\rangle = 1$ for all q'. Any ket $|P\rangle$ in P can be expressed as the result of a linear operator $\psi(q)$ acting on the standard ket,

$$|P\rangle = \psi(q)\rangle \tag{4.116}$$

the functional form of ψ being determined from the relation

$$\psi(q') = \langle q' \mid P\rangle \tag{4.117}$$

It is possible to choose the phase factors so that

$$p_{\alpha i}\,|P\rangle = -i\hbar\,\frac{\partial \psi}{\partial q_{\alpha i}}\Big\rangle \tag{4.118}$$

This additional requirement characterizes Schrödinger's representation, which for given canonical variables q and p is fixed apart from an arbitrary *constant* phase factor. Using Schrödinger's representation we may write the basis vectors for the product space S as $|\text{vac}; \psi_m\rangle$ and $|\boldsymbol{k}_1, \lambda_1; \ldots ; \boldsymbol{k}_n, \lambda_n ; \psi_m\rangle$, where $\langle q' \mid m\rangle = \psi_m(q')$. We then have

$$p_{\alpha i}\,|\text{vac}; \psi_m\rangle = \left|\text{vac}; -i\hbar\,\frac{\partial \psi_m}{\partial q_{\alpha i}}\right\rangle \tag{4.119}$$

with an analogous expression for $p_{\alpha i}\,|\boldsymbol{k}_1, \lambda_1; \ldots ; \boldsymbol{k}_n, \lambda_n ; \psi_m\rangle$.

PROBLEM

Show that $\boldsymbol{a}(\boldsymbol{x})$, $\boldsymbol{\pi}(\boldsymbol{x})$, $\boldsymbol{q}_\alpha$ and $\boldsymbol{p}_\alpha$, considered as operators on the product space S, form an irreducible set.

4.5 The free field

The equations of quantum electrodynamics can not be solved exactly except when the field is free, i.e. when there are no charged particles at all present. In this section we confine our attention to the free field. (Some approximate solutions to the equations of motion for the coupled systems will be given in Chapter 6.) The Hamiltonian H is now just the radiation Hamiltonian H_{rad} and the Hilbert space S on which H operates is just the photon space F. There is no need to consider a space P for charged particles, since these are assumed not to exist initially and since, in the non-relativistic approximation, their number is fixed. The Heisenberg equations for the creation and annihilation operators, and hence for the field operators, may be derived by using the commutation relations (4.44) and (4.45). For the annihilation operators we have

$$\begin{aligned}\dot{a}^{(\lambda)}(\boldsymbol{k}, t) &= \frac{1}{i\hbar}[a^{(\lambda)}(\boldsymbol{k}, t), H_{rad}] \\ &= -ikca^{(\lambda)}(\boldsymbol{k}, t) \qquad (4.120)\end{aligned}$$

If the Schrödinger and Heisenberg pictures are taken to coincide at $t=0$, then the Heisenberg annihilation operators are given by

$$a^{(\lambda)}(\boldsymbol{k}, t) = a^{(\lambda)}(\boldsymbol{k})e^{-ikct} \qquad (4.121)$$

where $a^{(\lambda)}(\boldsymbol{k})$ is a Schrödinger annihilation operator. The corresponding creation operators are related by the equation

$$a^{(\lambda)\dagger}(\boldsymbol{k}, t) = a^{(\lambda)\dagger}(\boldsymbol{k})e^{ikct} \qquad (4.122)$$

It follows that the Heisenberg operators for the transverse electric and magnetic induction fields are

$$\begin{aligned}\boldsymbol{e}^{\perp}(\boldsymbol{x}, t) = \frac{1}{2\pi}\sum_{\lambda}\iiint d^3k(\hbar ck)^{1/2}i\{&\hat{\boldsymbol{e}}^{(\lambda)}(\boldsymbol{k})a^{(\lambda)}(\boldsymbol{k})e^{i(\boldsymbol{k}.\boldsymbol{x}-kct)} \\ &-\hat{\boldsymbol{e}}^{(\lambda)*}(\boldsymbol{k})a^{(\lambda)\dagger}(\boldsymbol{k})e^{-i(\boldsymbol{k}.\boldsymbol{x}-kct)}\} \qquad (4.123)\end{aligned}$$

and

$$\begin{aligned}\boldsymbol{b}(\boldsymbol{x}, t) = \frac{1}{2\pi}\sum_{\lambda}\iiint d^3k(\hbar ck)^{1/2}i\{&\hat{\boldsymbol{k}}\times\hat{\boldsymbol{e}}^{(\lambda)}(\boldsymbol{k})a^{(\lambda)}(\boldsymbol{k})e^{i(\boldsymbol{k}.\boldsymbol{x}-kct)} \\ &-\hat{\boldsymbol{k}}\times\hat{\boldsymbol{e}}^{(\lambda)*}(\boldsymbol{k})a^{(\lambda)\dagger}(\boldsymbol{k})e^{-i(\boldsymbol{k}.\boldsymbol{x}-kct)}\} \qquad (4.124)\end{aligned}$$

respectively. Thus each plane wave component of the free field has a simple harmonic time dependence.

In the Schrödinger picture the state vector $|F(t)\rangle$ for the field varies with time according to Schrödinger's equation (4.7) with $H = H_{\mathrm{rad}}$. Since the Hamiltonian H_{rad} is time-independent, the solution of this equation is given formally by

$$|F(t)\rangle = e^{-iH_{\mathrm{rad}}t/\hbar}\,|F(0)\rangle \tag{4.125}$$

where the time evolution operator $\exp(-iH_{\mathrm{rad}}t/\hbar)$ is unitary, as H_{rad} is Hermitian. Now the vacuum state vector and all n-photon state vectors are eigenvectors of the free-field Hamiltonian H_{rad}. Hence they are also eigenvectors of $\exp(-iH_{\mathrm{rad}}t/\hbar)$ with eigenvalues $\exp(-iEt/\hbar)$, where E is an eigenvalue of H_{rad}. Thus the solution $|F(t)\rangle$ of the time-dependent Schrödinger equation for which $|F(0)\rangle = |F\rangle$ is, from Equations (4.70) and (4.125),

$$\begin{aligned} |F(t)\rangle &= |\mathrm{vac}\rangle\langle \mathrm{vac} \mid F\rangle \\ &+ \sum_{n=1}^{\infty} \frac{1}{n!} \iiint \mathrm{d}^3k_1 \ldots \iiint \mathrm{d}^3k_n \sum_{\lambda_1} \ldots \sum_{\lambda_n} e^{-i(k_1+\ldots+k_n)ct} \\ &\times |\boldsymbol{k}_1, \lambda_1; \ldots ; \boldsymbol{k}_n, \lambda_n\rangle\langle \boldsymbol{k}_1, \lambda_1; \ldots ; \boldsymbol{k}_n, \lambda_n \mid F\rangle \end{aligned} \tag{4.126}$$

The probability amplitude for the vacuum state or any n-photon state changes at most by a phase factor and the probabilities or probability densities are constant in time, so long as the system is left to itself. Moreover, the state vector remains normalized, if it is normalized initially. (The conservation of probability also follows directly from the unitary nature of the time evolution operator.) For the free field the number of photons is a constant of the motion. Thus, for example, if the state vector is initially in the vacuum subspace or in some n-photon subspace (all of which are eigenspaces of the photon number operator N), then it is never carried out of that subspace by any subsequent causal motion of the system. When the field is coupled to charged particles, however, the number of photons can change, through the absorption or emission of radiation.

COMMUTATION RELATIONS FOR THE HEISENBERG FIELDS

The equal-time commutation relations for the transverse electric and magnetic induction fields are valid whether or not the field is free (see Section 4.2). The non-equal-time commutation relations can be calculated exactly only for the free field, since only in this case can exact expressions for the Heisenberg operators $\boldsymbol{e}^{\perp}(\boldsymbol{x}, t)$ and $\boldsymbol{b}(\boldsymbol{x}, t)$ be obtained. In writing down the free-field commutation relations, it is useful to introduce the singular D-function defined through either of the equivalent formulae

$$D(\boldsymbol{x}, t) = \frac{1}{(2\pi)^3} \iiint d^3k e^{i\boldsymbol{k}.\boldsymbol{x}} \frac{\sin kct}{k}$$
$$= -\frac{1}{(2\pi)^3} \iiint d^3k \frac{\sin(\boldsymbol{k}.\boldsymbol{x} - kct)}{k} \tag{4.127}$$

the integrations being over all of $\boldsymbol{k}$-space. The following properties are immediate consequences of the definition:

$$\left(\nabla^2 - \frac{1}{c^2}\frac{\partial^2}{\partial t^2}\right) D(\boldsymbol{x}, t) = 0 \tag{4.128}$$

$$D(\boldsymbol{x}, 0) = 0 \tag{4.129}$$

and

$$\frac{1}{c}\frac{\partial}{\partial t} D(\boldsymbol{x}, t)\,|_{t=0} = \delta(\boldsymbol{x}) \tag{4.130}$$

Thus D is the solution of the homogeneous wave equation that vanishes at $t=0$ and whose time derivative at $t=0$ is $c\,\delta(\boldsymbol{x})$. We have also

$$D(-\boldsymbol{x}, t) = D(\boldsymbol{x}, t) \qquad D(\boldsymbol{x}, -t) = -D(\boldsymbol{x}, t) \tag{4.131}$$

i.e. D is an even function of $\boldsymbol{x}$ and an odd function of t. The angular integration in the first of the formulae (4.127) can be carried out to give

$$D(\boldsymbol{x}, t) = \frac{1}{2\pi^2 x} \int_0^\infty dk \sin kx \sin kct$$
$$= \frac{1}{4\pi x}\{\delta(x - ct) - \delta(x + ct)\} \tag{4.132}$$

which shows that D vanishes everywhere except on the light cone ($x = \pm ct$), where it has a delta-function singularity.

To calculate the commutator bracket of two components of the transverse electric field evaluated at possibly different times, we use the expansion (4.123) for $\boldsymbol{e}^\perp(\boldsymbol{x}, t)$, the commutation relations (4.44) and (4.45) for the creation and annihilation operators and the completeness relation (4.41) for the polarization vectors. The result is

$$[e_i^\perp(\boldsymbol{x}, t), e_j^\perp(\boldsymbol{x}', t')]$$
$$= 4\pi i\hbar c \frac{1}{(2\pi)^3} \iiint d^3k (k^2 \delta_{ij} - k_i k_j) \frac{\sin[\boldsymbol{k}.(\boldsymbol{x} - \boldsymbol{x}') - kc(t - t')]}{k}$$
$$= 4\pi i\hbar c \left(\frac{\partial^2}{\partial x_i\, \partial x_j'} - \delta_{ij} \frac{1}{c^2} \frac{\partial^2}{\partial t\, \partial t'}\right) D(\boldsymbol{x} - \boldsymbol{x}', t - t') \tag{4.133}$$

where the last line follows from the second of the formulae (4.127) for the D-function. In a similar manner, using also the expansion (4.124) for $\boldsymbol{b}(\boldsymbol{x}, t)$, we obtain

$$[e_i^{\perp}(\boldsymbol{x}, t), b_j(\boldsymbol{x}', t')] = 4\pi i\hbar c\varepsilon_{ijk}\frac{1}{c}\frac{\partial^2}{\partial t\,\partial x_k'}D(\boldsymbol{x}-\boldsymbol{x}', t-t') \qquad (4.134)$$

For the commutator of two components of the magnetic induction field we need the relation

$$\sum_{\lambda}[\hat{\boldsymbol{k}}\times\hat{\boldsymbol{e}}^{(\lambda)}(\boldsymbol{k})]_i[\hat{\boldsymbol{k}}\times\hat{\boldsymbol{e}}^{(\lambda)*}(\boldsymbol{k})]_j = \delta_{ij} - \hat{k}_i\hat{k}_j \qquad (4.135)$$

which follows from the orthogonality and completeness properties of the polarization vectors. It may then be shown that

$$[b_i(\boldsymbol{x}, t), b_j(\boldsymbol{x}', t')] = [e_i^{\perp}(\boldsymbol{x}, t), e_j^{\perp}(\boldsymbol{x}', t')] \qquad (4.136)$$

The equal-time commutation relations are special cases of the above, as may be verified by putting $t'=t$. Since D vanishes everywhere except on the light cone, the commutator brackets for the components of $\boldsymbol{e}^{\perp}$ and $\boldsymbol{b}$ are zero unless the space-time point $(\boldsymbol{x}', t')$ is infinitesimally close to the light cone with vertex at $(\boldsymbol{x}, t)$. Thus the measurement of two field strengths will not interfere when the space-time regions over which the measurements are taken can not be connected by light signals. This is in agreement with the relativistic principle of causality and with the local character† of $\boldsymbol{e}^{\perp}$ and $\boldsymbol{b}$. As was mentioned in Chapter 1, it is more meaningful to speak of measuring the average fields over a given space region V and a given time interval T, rather than the fields themselves at a point $\boldsymbol{x}$ and a time t. The average field operators may be defined by

$$E_i^{\perp}(V, T) = \frac{1}{VT}\iiint_V\int_T e_i^{\perp}(\boldsymbol{x}, t)\,\mathrm{d}^3x\,\mathrm{d}t \qquad (4.137)$$

and by

$$B_i(V, T) = \frac{1}{VT}\iiint_V\int_T b_i(\boldsymbol{x}, t)\,\mathrm{d}^3x\,\mathrm{d}t \qquad (4.138)$$

and the commutators for the average fields over two space-time regions VT and $V'T'$ may be expressed as integrals, taken over these regions, of

† It should be noted that the vector potential $\boldsymbol{a}$ in the Coulomb gauge is non-local in the sense that a measurement of $\boldsymbol{a}$ at $(\boldsymbol{x}, t)$ requires simultaneous measurements of $\boldsymbol{b}$, at time t, throughout all space. This follows from Equation (2.37) (which holds as well in the quantum as in the classical theory) and is reflected in the commutation relations for the components of $\boldsymbol{a}$ (see Problem (1) at the end of this section).

the commutators of the fields themselves. In certain simple cases the integrals are zero. Thus when the time intervals T and T' coincide, we have, since $D(\boldsymbol{x}, t)$ is an odd function of t,

$$[E_i^{\perp}(V, T), E_j^{\perp}(V', T)] = 0$$
$$= [B_i(V, T), B_j(V', T)] \tag{4.139}$$

for any space regions V and V', and when the space regions V and V' coincide, we have, since $D(\boldsymbol{x}, t)$ is an even function of $\boldsymbol{x}$,

$$[E_i^{\perp}(V, T), B_j(V, T')] = 0 \tag{4.140}$$

for any time intervals T and T'. If the space-time regions are such that some light signals emitted from V during T can reach V' during T', or vice versa, then the commutators are, in general, non-zero. The associated uncertainty relations are in agreement with the qualitative considerations of Chapter 1 and with the way in which the fields can, at least in principle, be measured by means of charged test bodies (Bohr and Rosenfeld 1933, 1950, Heitler 1954). Since the commutation relations derived here refer only to the free field, the charged particles composing the test bodies must be regarded as part of the macroscopic measuring apparatus and not as part of the quantum mechanical system being observed. If charged particles are included in the observed system, approximate commutation relations for the coupled field and particles must be used in deriving the uncertainty relations.

COHERENT STATES OF THE RADIATION FIELD

The n-photon states of the radiation field resemble incoherent superpositions of classical plane-wave states, since they are associated with definite wavevectors and polarization vectors but do not have well defined phases (in the classical sense). Thus the expectation value of the transverse electric or magnetic induction field (in the Heisenberg picture) for any n-photon state is zero, whereas a classical plane-wave field has a simple-harmonic time dependence. There exist states of the quantized field, called coherent or quasi-classical states (Glauber 1963), such that the phase is more well defined but the number of photons, and hence the energy and momentum, is less sharp than for the n-photon states. More particularly, consider the following solution of the classical free-field equations for the transverse vector potential:

$$\mathbf{A}(\boldsymbol{x}, t) = \frac{1}{2\pi}\sum_{\lambda}\iiint \mathrm{d}^3k\left(\frac{\hbar c}{k}\right)^{1/2}\{\hat{\boldsymbol{e}}^{(\lambda)}(\boldsymbol{k})\alpha^{(\lambda)}(\boldsymbol{k})e^{i(\boldsymbol{k}.\boldsymbol{x}-kct)}$$
$$+\hat{\boldsymbol{e}}^{(\lambda)*}(\boldsymbol{k})\alpha^{(\lambda)*}(\boldsymbol{k})e^{-i(\boldsymbol{k}.\boldsymbol{x}-kct)}\} \tag{4.141}$$

where the $\alpha^{(\lambda)}(\boldsymbol{k})$ are c-number amplitudes related to the Fourier transform of $\mathbf{A}$ in the same way as the annihilation operators $a^{(\lambda)}(\boldsymbol{k})$ are related to the Fourier transform of $\boldsymbol{a}$ (see Section 4.3). It is assumed that the classical energy, which is given by

$$E = \frac{1}{8\pi}\iiint\{\boldsymbol{E}^{\perp 2} + \boldsymbol{B}^2\}\,\mathrm{d}^3x = \frac{1}{8\pi}\iiint\left\{\frac{1}{c^2}\dot{\mathbf{A}}^2 + (\nabla\times\mathbf{A})^2\right\}\mathrm{d}^3x$$
$$= \sum_\lambda\iiint \mathrm{d}^3k\,k\hbar c k\alpha^{(\lambda)*}(\boldsymbol{k})\alpha^{(\lambda)}(\boldsymbol{k}) \tag{4.142}$$

is finite. Then the coherent state α of the quantized field corresponding to the classical state described by Equation (4.141) is represented by a vector $|\alpha\rangle$ normalized to unity and such that (i) the expectation value of the vector potential operator (in the Heisenberg picture) is the classical vector potential,

$$\langle\alpha|\,\boldsymbol{a}(\boldsymbol{x}, t)\,|\alpha\rangle = \mathbf{A}(\boldsymbol{x}, t) \tag{4.143}$$

and (ii) the expectation value of the Hamiltonian is the classical energy,

$$\langle\alpha|\,H_{\mathrm{rad}}\,|\alpha\rangle = E \tag{4.144}$$

Such a state exists and is uniquely determined by properties (i) and (ii) (Kroll 1964).

We first introduce the amplitude displacement operator $D(\alpha)$ defined by

$$D(\alpha) = e^{\alpha a^\dagger - \alpha^* a} \tag{4.145}$$

where, in an obviously consistent notation,

$$\alpha a^\dagger = \sum_\lambda\iiint \mathrm{d}^3k\,\alpha^{(\lambda)}(\boldsymbol{k})a^{(\lambda)\dagger}(\boldsymbol{k}) \tag{4.146}$$

and

$$\alpha^* a = \sum_\lambda\iiint \mathrm{d}^3k\,\alpha^{(\lambda)*}(\boldsymbol{k})a^{(\lambda)}(\boldsymbol{k}) \tag{4.147}$$

We also put

$$|\alpha|^2 = \sum_\lambda\iiint \mathrm{d}^3k\,|\alpha^{(\lambda)}(\boldsymbol{k})|^2 \tag{4.148}$$

The following commutation relations may be verified by using the commutation relations for the creation and annihilation operators:

$$[a^{(\lambda)}(\boldsymbol{k}), \alpha a^\dagger] = \alpha^{(\lambda)}(\boldsymbol{k}) \tag{4.149}$$

$$[a^{(\lambda)\dagger}(\boldsymbol{k}), \alpha^* a] = -\alpha^{(\lambda)*}(\boldsymbol{k}) \tag{4.150}$$

and

$$[\alpha^* a, \alpha a^\dagger] = |\alpha|^2 \tag{4.151}$$

Moreover, all the annihilation operators commute with $\alpha^* a$ and all the creation operators with $\alpha a^\dagger$. Since $\alpha a^\dagger - \alpha^* a$ is antihermitian, $D(\alpha)$ is unitary. Using Theorem 2 of Appendix D and the commutation relation (4.151) we may write $D(\alpha)$ in normally ordered form, with all creation operators appearing to the left of all annihilation operators. Thus

$$D(\alpha) = e^{-\frac{1}{2}|\alpha|^2} e^{\alpha a^\dagger} e^{-\alpha^* a} \tag{4.152}$$

We have also from Theorem 1 of Appendix D

$$\begin{aligned} D^\dagger(\alpha) a^{(\lambda)}(\boldsymbol{k}) D(\alpha) &= a^{(\lambda)}(\boldsymbol{k}) + [\alpha^* a - \alpha a^\dagger, a^{(\lambda)}(\boldsymbol{k})] \\ &= a^{(\lambda)}(\boldsymbol{k}) + \alpha^{(\lambda)}(\boldsymbol{k}) \end{aligned} \tag{4.153}$$

which justifies the term amplitude displacement operator. The state vector $|\alpha\rangle$ is now defined to be the vacuum state vector displaced by $D(\alpha)$, i.e.

$$\begin{aligned} |\alpha\rangle &= D(\alpha)\,|\text{vac}\rangle \\ &= e^{-\frac{1}{2}|\alpha|^2} e^{\alpha a^\dagger}\,|\text{vac}\rangle \end{aligned} \tag{4.154}$$

where the last expression is obtained by using the normally ordered form of $D(\alpha)$ and the fact that the vacuum state, since it is annihilated by $\alpha^* a$, is an eigenstate of $\exp(-\alpha^* a)$ with eigenvalue 1. Because $D(\alpha)$ is unitary and the vector $|\text{vac}\rangle$ is normalized to unity, the vector $|\alpha\rangle$ is also normalized to unity. Furthermore, because $D(\alpha)$ is an amplitude displacement operator,

$$D^\dagger(\alpha) H_{\text{rad}} D(\alpha) = \sum_\lambda \iiint \mathrm{d}^3k \hbar ck \{a^{(\lambda)\dagger}(\boldsymbol{k}) + \alpha^{(\lambda)*}(\boldsymbol{k})\}\{a^{(\lambda)}(\boldsymbol{k}) + \alpha^{(\lambda)}(\boldsymbol{k})\} \tag{4.155}$$

and thus

$$\begin{aligned} \langle\alpha|\, H_{\text{rad}}\,|\alpha\rangle &= \langle\text{vac}|\, D^\dagger(\alpha) H_{\text{rad}} D(\alpha)\,|\text{vac}\rangle \\ &= \sum_\lambda \iiint \mathrm{d}^3k \hbar ck\, |\alpha^{(\lambda)}(\boldsymbol{k})|^2 \\ &= E \end{aligned} \tag{4.156}$$

Here the expansion (4.101) for the radiation Hamiltonian and the relations $a^{(\lambda)}(\boldsymbol{k})\,|\text{vac}\rangle = 0$ and $0 = \langle\text{vac}|\, a^{(\lambda)\dagger}(\boldsymbol{k})$ have also been used. In a similar way, using the expansion (4.46) for the vector potential operator, we obtain

$$\langle\alpha|\, \boldsymbol{a}(\boldsymbol{x}, t)\,|\alpha\rangle = \mathbf{A}(\boldsymbol{x}, t) \tag{4.157}$$

which implies, incidentally, that

$$\langle\alpha|\, \boldsymbol{e}^{\perp}(\boldsymbol{x}, t)\, |\alpha\rangle = \boldsymbol{E}^{\perp}(\boldsymbol{x}, t) \tag{4.158}$$

and

$$\langle\alpha|\, \boldsymbol{b}(\boldsymbol{x}, t)\, |\alpha\rangle = \boldsymbol{B}(\boldsymbol{x}, t). \tag{4.159}$$

The state α defined through Equation (4.154) thus satisfies all the requirements for a coherent state. To prove the uniqueness, we let $|\alpha\rangle$ be any normalized vector having the properties (i) and (ii) expressed by Equations (4.143) and (4.144) and let $D(\alpha)$ be defined, as before, by Equation (4.145). From property (i) and the relation of $a^{(\lambda)}(\boldsymbol{k})$ and $\alpha^{(\lambda)}(\boldsymbol{k})$ to the Fourier transforms of $\boldsymbol{a}$ and $\mathbf{A}$ we obtain

$$\langle\alpha|\, a^{(\lambda)}(\boldsymbol{k})\, |\alpha\rangle = \alpha^{(\lambda)}(\boldsymbol{k}) \tag{4.160}$$

for all $\boldsymbol{k}$, λ. If $|F\rangle = D^{\dagger}(\alpha)\,|\alpha\rangle$, then $|F\rangle$ is normalized to unity, since $D(\alpha)$ is unitary and $|\alpha\rangle$ is normalized to unity. Also, since $D^{\dagger}(\alpha)$ is an inverse amplitude displacement operator,

$$\begin{aligned}\langle F|\, a^{(\lambda)}(\boldsymbol{k})\, |F\rangle &= \langle\alpha|\, D(\alpha)a^{(\lambda)}(\boldsymbol{k})D^{\dagger}(\alpha)\, |\alpha\rangle \\ &= \langle\alpha|\, \{a^{(\lambda)}(\boldsymbol{k}) - \alpha^{(\lambda)}(\boldsymbol{k})\}\, |\alpha\rangle \\ &= 0\end{aligned} \tag{4.161}$$

as follows from Equation (4.160). Equation (4.161) is satisfied when $|F\rangle$ is the vacuum state vector or any normalized vector in an n-photon subspace. Property (ii), however, holds only if

$$\begin{aligned}E &= \langle F|\, D^{\dagger}(\alpha)H_{\text{rad}}D(\alpha)\, |F\rangle \\ &= \sum_{\lambda} \iiint \mathrm{d}^3k\hbar ck \langle F|\, a^{(\lambda)\dagger}(\boldsymbol{k})a^{(\lambda)}(\boldsymbol{k})\, |F\rangle \\ &\quad + \sum_{\lambda} \iiint \mathrm{d}^3k\hbar ck\, |\alpha^{(\lambda)}(\boldsymbol{k})|^2\end{aligned} \tag{4.162}$$

the cross terms vanishing because of Equation (4.161). The second term on the right-hand side of Equation (4.162) is just E, and so

$$\sum_{\lambda} \iiint \mathrm{d}^3k\hbar ck \langle F|\, a^{(\lambda)\dagger}(\boldsymbol{k})a^{(\lambda)}(\boldsymbol{k})\, |F\rangle = 0 \tag{4.163}$$

Since the integrand is positive definite, the equation

$$a^{(\lambda)}(\boldsymbol{k})\, |F\rangle = 0 \tag{4.164}$$

must hold almost everywhere in $\boldsymbol{k}$-space, for each value of λ. The

n-photon amplitudes $\langle \text{vac}|\, a^{(\lambda_1)}(\boldsymbol{k}_1) \ldots a^{(\lambda_n)}(\boldsymbol{k}_n)\,|F\rangle$ therefore vanish almost everywhere† and hence make no contribution in the expansion (4.70) for $|F\rangle$. Thus $|F\rangle$, as it is normalized to unity, must differ from the vacuum state vector by at most a phase factor. It follows that $|\alpha\rangle$ is defined to within a phase factor and that the corresponding state α is uniquely determined.

PROBLEMS

(1) Prove that the components of the free-field vector potential in the Coulomb gauge satisfy the commutation relation

$$[a_i(\boldsymbol{x}, t), a_j(\boldsymbol{x}', t')] = 4\pi i\hbar c\left\{\frac{\partial^2}{\partial x_i\, \partial x_j'} H(\boldsymbol{x}-\boldsymbol{x}', t-t') - \delta_{ij} D(\boldsymbol{x}-\boldsymbol{x}', t-t')\right\}$$

where‡

$$H(\boldsymbol{x}, t) = \frac{1}{(2\pi)^3} \iiint \mathrm{d}^3k\, e^{i\boldsymbol{k}.\boldsymbol{x}} \frac{\sin kct}{k^3}$$

Derive the properties

$$\nabla^2 H = -D(\boldsymbol{x}, t) = \frac{1}{c^2}\frac{\partial^2 H}{\partial t^2} \qquad H(\boldsymbol{x}, 0) = 0$$

and use contour integration to show that

$$H(\boldsymbol{x}, t) = \begin{cases} \dfrac{1}{4\pi}\varepsilon(t) & 0 \leqslant x \leqslant c\,|t| \\ \dfrac{1}{4\pi}\dfrac{ct}{x} & x > c\,|t| \end{cases}$$

where $\varepsilon(t) = 1$, 0 or -1 according as $t >$, $=$, or <0. Show also that if $\boldsymbol{x} \neq 0$, then

$$H(\boldsymbol{x}, t) = \frac{1}{8\pi x}(|x+ct| - |x-ct|)$$

Note that H is constant inside both the past and the future light cone, but has non-zero derivatives outside the light cone.

† The n-photon amplitudes are identically zero if they are assumed to be continuous functions of $\boldsymbol{k}_1, \ldots, \boldsymbol{k}_n$ for each set of values of $\lambda_1, \ldots, \lambda_n$.

‡ The function H defined here differs in sign from that denoted by the same symbol by Heitler (1954).

(2) Use the commutation relations for the components of $\boldsymbol{a}$ to deduce (i) the commutation relations for the components of $\boldsymbol{e}^{\perp}$ and $\boldsymbol{b}$ and (ii) the equal-time canonical commutation relations.

(3) Show that the coherent state α is a common eigenstate of the annihilation operators $a^{(\lambda)}(\boldsymbol{k})$ with corresponding eigenvalues $\alpha^{(\lambda)}(\boldsymbol{k})$. Show also that the vacuum probability for this state is $\exp(-|\alpha|^2)$ and hence that the n-photon probability density is $\exp(-|\alpha|^2)\,|\alpha^{(\lambda_1)}(\boldsymbol{k}_1)|^2 \ldots |\alpha^{(\lambda_n)}(\boldsymbol{k}_n)|^2$. Deduce that the vacuum state is the only coherent state which is also an eigenstate of H_{rad}.

(4) Show that in the Schrödinger picture the coherent state α corresponds to the vector $|\alpha(t)\rangle = D(\alpha, t)\,|\text{vac}\rangle$, where the time-dependent displacement operator $D(\alpha, t)$ is obtained from $D(\alpha)$ by replacing $\alpha^{(\lambda)}(\boldsymbol{k})$ by $\alpha^{(\lambda)}(\boldsymbol{k})e^{-ikct}$. It is assumed that the Schrödinger and Heisenberg pictures coincide at $t=0$.

REFERENCES

Bohr, N. and Rosenfeld, L. (1933). "Zur Frage der Messbarkeit der Elecktromagnetischen Feldgrössen." *Det. Kgl. Danske Videnskabernes Selskab. Mathematisk-fysiske Meddelelser*, **12,** No. 8.

Bohr, N. and Rosenfeld, L. (1950). "Field and charge measurements in quantum electrodynamics." *Phys. Rev.* **78,** 794.

Dirac, P. A. M. (1925). "The fundamental equations of quantum mechanics." *Proc. R. Soc. Lond.* A **109,** 642.

Dirac, P. A. M. (1958). "The Principles of Quantum Mechanics." Clarendon Press, Oxford.

Glauber, R. J. (1963). "Coherent and incoherent states of the radiation field." *Phys. Rev.* **131,** 2766.

Heisenberg, W. (1930). "The Physical Principles of the Quantum Theory." University of Chicago Press.

Heitler, W. (1954). "The Quantum Theory of Radiation." Clarendon Press, Oxford.

Kroll, N. (1964). *In* "Quantum Optics and Electronics" (ed. C. De Witt, A. Blandin and C. Cohen-Tannoudji). Gordon & Breach, New York.

von Neumann, J. (1931). "Die Eindeutigkeit der Schrödingerschen Operatoren." *Math. Ann.* **104,** 570.

von Neumann, J. (1955). "Mathematical Foundations of Quantum Mechanics." Princeton University Press.

Wightman, A. S. and Schweber, S. S. (1955). "Configuration space methods in relativistic quantum field theory. I." *Phys. Rev.* **98,** 812.

Chapter 5

Symmetries and Conservation Laws

5.1 Relations between observers

The connection between symmetries and conservation laws has been mentioned previously in the context of the Lagrangian formulation of classical electrodynamics. A similar connection exists in quantum electrodynamics and to establish it we examine the relation between the descriptions of the system as given by two observers, O and $\bar{\text{O}}$, say. We assume that each observer uses a rectangular Cartesian reference frame of specified origin, handedness and orientation and that each uses a definite convention for the origin and sense of time and for the signs of the charged particles. The reference frames are assumed also to be inertial frames and to be such that the non-relativistic approximation is valid for both observers. More particularly, the transformation between O and $\bar{\text{O}}$ will be restricted to be a time displacement, space displacement, space rotation, charge conjugation, space inversion, time reversal or a combination of any or all of these. This implies that the two observers agree on the units of length, time and charge; we suppose them also to agree on the unit of mass. The transformations listed above are those under which the combined Maxwell–Lorentz and Newtonian equations are covariant and therefore define symmetries of the system. It is sometimes convenient to suppose that there are prescribed external static fields acting on the charged particles. These may destroy some of the symmetries, but even so it is still possible to relate the descriptions given by O and $\bar{\text{O}}$. It should be noted also that we are adopting, as before, the alias or passive point of view—there is one physical system and two observers.

To relate the two descriptions of the system we work in the Schrödinger picture and use the following conventions:

(1) The two observers ascribe to states of the system rays in a common Hilbert space S.

(2) The two observers employ the same irreducible set of operators on S and, relative to their own reference systems, interpret them in the same

way. Thus if O uses the operator $q_{\alpha i}$ to refer to the displacement of particle α from his origin along his ith axis, then $\bar{\text{O}}$ uses the *same* operator to refer to the displacement of the same particle from *his* origin along *his* ith axis. Similar considerations apply to the momentum operators $p_{\alpha i}$. Also, $e_i^{\perp}(\boldsymbol{x})$ and $b_i(\boldsymbol{x})$ are the ith components of the field operators at the point with coordinates $\boldsymbol{x}$ for *both* observers. The column vector $\boldsymbol{x}$, or $(x_1, x_2, x_3)^{\mathrm{T}}$, is associated with a point P by O and with a point $\bar{P}$ by $\bar{\text{O}}$, where $\bar{P}$ bears the same relation to $\bar{\text{O}}$'s frame that P does to O's. (Cf. the discussion of rotations in connection with Noether's principle in Chapter 3.)

(3) The change in description on going from one observer to the other is effected by means of a change in the state vectors or rays assigned by the observers to the physical states. Thus if O ascribes the unit ray $\boldsymbol{S}(\boldsymbol{t})$ to a state of the system at some time, labelled t by O, then $\bar{\text{O}}$ ascribes a unit ray $\bar{\boldsymbol{S}}(\bar{t})$ to the same state at the same objective time,† labelled $\bar{t}$ by $\bar{\text{O}}$. The mapping $\boldsymbol{S}(t) \to \bar{\boldsymbol{S}}(\bar{t})$ from the set of unit rays of S into itself is supposed to be one-one and onto. This is because there is supposed to exist at any time, and for each observer, a one-to-one correspondence between physically realizable states and unit rays.

It should be emphasized that many other conventions could be used. Thus each observer could have his own Hilbert space, S for O and $\bar{S}$ for $\bar{\text{O}}$, say. (The spaces S and $\bar{S}$ would, however, have to be isomorphic.) Also the change in description could be carried by the operators instead of by the state vectors (analogously to the use of the Heisenberg rather than the Schrödinger picture) or, in an infinity of different ways, partly by the state vectors and partly by the operators. For any observer, the quantities of physical interest (eigenvalues, transition probabilities, etc.) are independent of the convention used.

Given the relation between the two observers, we can determine the transformation (in the Hilbert space S) from the following principles:

A. *Transition probabilities are independent of the observer*

If the system is in state 1, the probability that after a measurement is performed it is found to be in state 2 is independent of the observer. This means that if O ascribes the unit rays $\boldsymbol{S}_1(t)$ and $\boldsymbol{S}_2(t)$ to states 1 and 2 of the system at a time labelled t by him and $\bar{\text{O}}$ ascribes the unit rays $\bar{\boldsymbol{S}}_1(\bar{t})$ and $\bar{\boldsymbol{S}}_2(\bar{t})$ to the same states at the same objective time, labelled $\bar{t}$ by him, then

$$|\langle S_1(t) \mid S_2(t)\rangle|^2 = |\langle \bar{S}_1(\bar{t}) \mid \bar{S}_2(\bar{t})\rangle|^2 \tag{5.1}$$

† It is only in a non-relativistic theory that we can speak of an objective time. In a relativistic theory the state vectors would have to be labelled by a time-like parameter.

The two transition probabilities refer to one measurement on the system performed by either observer at a given instant of time. Thus $\bar{\text{O}}$ would use the same apparatus to measure the probability for $\bar{\boldsymbol{S}}_1 \rightarrow \bar{\boldsymbol{S}}_2$ at his time $\bar{t}$ that O uses to measure that for $\boldsymbol{S}_1 \rightarrow \boldsymbol{S}_2$ at his time t, although he would describe it differently (i.e. the apparatus would be in a different relation to $\bar{\text{O}}$'s frame and conventions than it would be to O's), and either observer would ascribe the reduction of his wave packet to letting the system interact with this apparatus and recording the result.

B. The correspondence principle

This implies that the expectation values of operators transform exactly like the corresponding classical dynamical variables. (The system under consideration has a classical analogue describable in terms of canonical coordinates and momenta.)

According to a theorem of Wigner (Wigner 1959, see also Fonda and Ghirardi 1970), the unit ray mapping $\boldsymbol{S}(t) \rightarrow \bar{\boldsymbol{S}}(\bar{t})$, which preserves transition probabilities, can be achieved by a vector mapping U from S onto itself, with U being either a linear or an antilinear† unitary operator that is unique apart from an arbitrary multiplicative phase factor. In all cases to be considered, the operator U is independent of the time and depends only on the observers O and $\bar{\text{O}}$ (indeed, only on the relation between O and $\bar{\text{O}}$). If the transformation $\text{O} \rightarrow \bar{\text{O}}$ is continuously connected to the identity (time translation, space translation, space rotation), then U must be linear, since the identity transformation is linear and the operators of a continuous set cannot suddenly change from being linear to being antilinear. Whether U is linear or antilinear may be determined anyway from the correspondence principle and the canonical commutation relations; it will be seen that U is antilinear only for the time-reversal transformation, or for product transformations involving an odd number of time reversals. We note that U is defined to within a phase factor by its effect on the operators Ω of an irreducible set. For if U' is any unitary operator such that for all Ω

$$U'^{\dagger}\Omega U' = U^{\dagger}\Omega U, \tag{5.2}$$

then $U'U^{\dagger}$ commutes with an irreducible set of operators and is therefore a multiple of the identity, i.e. U' is a constant times U. Moreover, the constant must have modulus one, since both U and U' are unitary.

For any space-time reference frame in which the classical Maxwell–Lorentz and Newtonian equations are valid the canonical formalism may

† For the definition of antilinear operators, see Section 5.2.

be set up and canonical quantization carried out. We may then pass, in the usual way, from the Heisenberg to the Schrödinger picture and, by suitable choice of the phase of the state vector as the state develops in time (Fonda and Ghirardi 1970), derive the Schrödinger equation for the system. This procedure may be followed by two observers O and Ō. Their respective Schrödinger equations are

$$i\hbar \frac{\mathrm{d}}{\mathrm{d}t} |S(t)\rangle = H\, |S(t)\rangle \tag{5.3}$$

and

$$i\hbar \frac{\mathrm{d}}{\mathrm{d}\bar{t}} |\bar{S}(\bar{t})\rangle = \bar{H}\, |\bar{S}(\bar{t})\rangle \tag{5.4}$$

where H is the Hamiltonian operator for O, $\bar{H}$ is that for Ō and where

$$|\bar{S}(\bar{t})\rangle = U(\mathrm{O} \to \bar{\mathrm{O}})\, |S(t)\rangle \tag{5.5}$$

It is assumed that the external fields, if any are present, have no time dependence in either frame, so that both H and $\bar{H}$ are time-independent. (For all transformations O → Ō to be considered, the external fields are time-independent for Ō if they are time-independent for O.) Since U is time-independent, substitution for $|\bar{S}(\bar{t})\rangle$ in the Schrödinger equation for Ō gives

$$i\hbar \frac{\mathrm{d}}{\mathrm{d}\bar{t}} |S(t)\rangle = \pm U^{\dagger}\bar{H}U\, |S(t)\rangle \tag{5.6}$$

where the plus or the minus sign holds according as U is linear or antilinear. Now $\mathrm{d}/\mathrm{d}\bar{t} = \pm \mathrm{d}/\mathrm{d}t$ with the minus sign being applicable only if there is time reversal, i.e. only if U is antilinear. Since $|S(t)\rangle$ is arbitrary, we obtain in all cases, by comparison of Equations (5.3) and (5.6),

$$\bar{H} = UHU^{\dagger} \tag{5.7}$$

This equation gives the relation between the two Hamiltonian operators, and includes the transformation of the external fields. If $\bar{H} = H$, the Heisenberg equations are the same for the two observers and the transformation O → Ō defines a symmetry of the system. A solution of the Schrödinger equation for O is then also a solution for Ō and for every (time-dependent) state described in a certain way by O there is a corresponding state described in exactly the same way by Ō. It follows from Equation (5.7) that the condition for a symmetry transformation is that the unitary operator U should commute with H.† The invariance of

† The charge conjugation transformation in non-relativistic quantum electrodynamics, to be discussed in Section 5.4, is an exception to this rule.

H as an operator in Hilbert space implies the invariance of its functional form also, since the two observers use the same operators for the canonical dynamical variables. Changes in the canonical dynamical variables themselves and the associated change in the functional form of the Hamiltonian (i.e. canonical transformations) will be considered in Chapter 7.

5.2 Linear and antilinear operators

An operator L on kets is linear if $L(c_1|S_1\rangle + c_2|S_2\rangle) = c_1L|S_1\rangle + c_2L|S_2\rangle$ for all kets $|S_1\rangle$, $|S_2\rangle$ and all complex numbers c_1, c_2. Let us recall how a linear operator on kets defines a linear operator on bras (Dirac 1958). Bra vectors are linear functionals on the space of ket vectors and there is a one-to-one correspondence between the bras and the kets such that $\langle S_1| c_1^* + \langle S_2| c_2^*$ corresponds to $c_1|S_1\rangle + c_2|S_2\rangle$. For fixed $\langle R|$ the number $\langle R|(L|S\rangle)$ depends linearly on $|S\rangle$ and hence equals $\langle R' | S\rangle$ for some bra $\langle R'|$. $\langle R'|$ depends linearly on $\langle R|$ and is thus the result of a linear operator on bras applied to $\langle R|$. This operator is uniquely determined by L and is also denoted by L, so that $\langle R'| = \langle R| L$. Since then $(\langle R| L)|S\rangle = \langle R|(L|S\rangle)$, the brackets may be omitted and the matrix element written as $\langle R| L |S\rangle$ with L operating either to the right on $|S\rangle$ or to the left on $\langle R|$. An operator A on kets is *antilinear* if $A(c_1|S_1\rangle + c_2|S_2\rangle) = c_1^*A|S_1\rangle + c_2^*A|S_2\rangle$ for all kets $|S_1\rangle$, $|S_2\rangle$ and all complex numbers c_1, c_2. An antilinear operator on kets may be used to define an antilinear operator on bras. For fixed $\langle R|$ the number $\{\langle R|(A|S\rangle)\}^*$ depends linearly on $|S\rangle$ and hence equals $\langle R' | S\rangle$ for some bra $\langle R'|$. $\langle R'|$ depends antilinearly on $\langle R|$ and is thus the result of an antilinear operator on bras applied to $\langle R|$. This operator is uniquely determined by A and may also be denoted by A, so that $\langle R'| = \langle R| A$. It follows that $(\langle R| A)|S\rangle = \{\langle R|(A|S\rangle)\}^*$. Now, however, if the brackets are to be removed, we must distinguish between A's operating to the left and to the right. Denoting the operator on bras by $\overleftarrow{A}$ and the corresponding operator on kets by $\vec{A}$, we have $\langle R| \overleftarrow{A} |S\rangle = \langle R| \vec{A} |S\rangle^*$. Thus corresponding matrix elements of $\overleftarrow{A}$ and $\vec{A}$ are complex conjugates of each other.

The Hermitian adjoint of a linear operator L is defined as follows. The ket $|S'\rangle$ that corresponds to the bra $\langle S| L$ depends linearly on the ket $|S\rangle$. (It depends antilinearly on the bra $\langle S|$.) Hence $|S'\rangle$ is the result of a linear operator applied to $|S\rangle$. This linear operator is the Hermitian adjoint of L and is denoted by $L^\dagger$. Since $\langle R| L^\dagger |S\rangle = \langle S| L |R\rangle^*$ for any $|R\rangle$ and $|S\rangle$, we have $\langle R|(L^\dagger)^\dagger |S\rangle = \langle S| L^\dagger |R\rangle^* = \langle R| L |S\rangle$ and thus $(L^\dagger)^\dagger = L$. Also $\langle R| L_2^\dagger L_1^\dagger |S\rangle = \langle S| L_1L_2 |R\rangle^* = \langle R|(L_1L_2)^\dagger |S\rangle$ and thus $(L_1L_2)^\dagger = L_2^\dagger L_1^\dagger$.

An analogous procedure may be carried out for antilinear operators. The ket $|S'\rangle$ that corresponds to the bra $\langle S|\,\overleftarrow{A}$ depends antilinearly on the ket $|S\rangle$ and is therefore the result of an antilinear operator $\vec{A}^{\dagger}$, the Hermitian adjoint of $\vec{A}$, applied to $|S\rangle$. It follows that $\langle R|\,\vec{A}^{\dagger}\,|S\rangle = \langle S|\,\overleftarrow{A}\,|R\rangle^* = \langle S|\,\vec{A}\,|R\rangle$, in contrast to the relation between the matrix elements of L and $L^{\dagger}$. The Hermitian adjoint $\overleftarrow{A}^{\dagger}$ of the antilinear operator $\overleftarrow{A}$ on bras may be defined in a similar way—$\overleftarrow{A}^{\dagger}$ is also antilinear and such that $\langle R|\,\overleftarrow{A}^{\dagger}\,|S\rangle = \langle S|\,\vec{A}\,|R\rangle^* = \langle S|\,\overleftarrow{A}\,|R\rangle$. Moreover, $\overleftarrow{A}^{\dagger}$ is the antilinear operator on bras corresponding to the antilinear operator $\vec{A}^{\dagger}$ on kets, since $\langle R|\,\overleftarrow{A}^{\dagger}\,|S\rangle = \langle R|\,\vec{A}^{\dagger}\,|S\rangle^*$. It may also be shown that $(\vec{A}^{\dagger})^{\dagger} = \vec{A}$ and $(\overleftarrow{A}^{\dagger})^{\dagger} = \overleftarrow{A}$. The product $\vec{A}_1\vec{A}_2$ of two antilinear operators $\vec{A}_1$ and $\vec{A}_2$ is linear. Similarly $\overleftarrow{A}_1\overleftarrow{A}_2$, operating on bras, is linear. Now linear operators on bras or kets define linear operators on kets or bras, respectively. The linear operator on bras defined by $\vec{A}_1\vec{A}_2$ is $\overleftarrow{A}_1\overleftarrow{A}_2$ and that on kets defined by $\overleftarrow{A}_1\overleftarrow{A}_2$ is $\vec{A}_1\vec{A}_2$, as follows from $\langle R|\,\overleftarrow{A}_1\overleftarrow{A}_2\,|S\rangle = \langle R|\,\overleftarrow{A}_1\vec{A}_2\,|S\rangle^* = \langle R|\,\vec{A}_1\vec{A}_2\,|S\rangle$. Also $(\vec{A}_1\vec{A}_2)^{\dagger}$, in so far as it is a linear operator on kets, equals $\vec{A}_2^{\dagger}\vec{A}_1^{\dagger}$. Similarly $(\overleftarrow{A}_1\overleftarrow{A}_2)^{\dagger}$ and $\overleftarrow{A}_2^{\dagger}\overleftarrow{A}_1^{\dagger}$ are equal, as operators on bras. If L is linear and $\vec{A}$ antilinear, then $L\vec{A}$ and $\vec{A}L$ are antilinear operators on kets. The corresponding antilinear operators on bras are $L\overleftarrow{A}$ and $\overleftarrow{A}L$, respectively. (L can operate to the left on a bra or to the right on a ket.) Also $(L\vec{A})^{\dagger} = \vec{A}^{\dagger}L^{\dagger}$ and $(\overleftarrow{A}L)^{\dagger} = L^{\dagger}\overleftarrow{A}^{\dagger}$. Thus the rule $(\Omega_1\Omega_2)^{\dagger} = \Omega_2^{\dagger}\Omega_1^{\dagger}$ holds for any linear or antilinear operators Ω_1, Ω_2, provided the products are defined.

A linear operator U is unitary if $UU^{\dagger} = 1 = U^{\dagger}U$. The inverse of a unitary operator is unitary and the product of two unitary operators is unitary. Putting $|\bar{R}\rangle = U\,|R\rangle$ and $|\bar{S}\rangle = U\,|S\rangle$ we obtain $\langle\bar{R}\mid\bar{S}\rangle = \langle R\mid S\rangle$, i.e. U preserves scalar products and hence norms. U also preserves Hermiticity, since if Ω is a Hermitian linear or antilinear operator (in the sense that $\Omega = \Omega^{\dagger}$), then so is the linear or antilinear operator $\bar{\Omega}$ defined by $\bar{\Omega} = U\Omega U^{\dagger}$. The operators Ω and $\bar{\Omega}$ (not necessarily Hermitian or even linear or antilinear) have the same eigenvalue spectrum, since $\Omega\,|S\rangle = c\,|S\rangle$ implies $\bar{\Omega}\,|\bar{S}\rangle = c\,|\bar{S}\rangle$, and conversely. If f is a function of operators $\Omega_1, \Omega_2, \ldots$, then $\overline{Uf(\Omega_1, \Omega_2, \ldots)U^{\dagger}} = f(\bar{\Omega}_1, \bar{\Omega}_2, \ldots)$, as follows from the relations $(\overline{c\Omega}) = c\bar{\Omega}$, $(\overline{\Omega_1 + \Omega_2}) = \bar{\Omega}_1 + \bar{\Omega}_2$ and $(\overline{\Omega_1\Omega_2}) = \bar{\Omega}_1\bar{\Omega}_2$. An antilinear operator $\vec{U}$ is said to be unitary if $\vec{U}\vec{U}^{\dagger} = 1 = \vec{U}^{\dagger}\vec{U}$, where 1 is the identity operator on kets. Since $\vec{U}\vec{U}^{\dagger}$ and $\vec{U}^{\dagger}\vec{U}$ are linear, $\overleftarrow{U}\overleftarrow{U}^{\dagger} = 1 = \overleftarrow{U}^{\dagger}\overleftarrow{U}$, where 1 is the identity operator on bras, and so, in this sense, $\overleftarrow{U}$ is also unitary. Putting $|\bar{R}\rangle = \vec{U}\,|R\rangle$ and $|\bar{S}\rangle = \vec{U}\,|S\rangle$ we obtain $\langle\bar{R}\mid\bar{S}\rangle = \langle R\mid S\rangle^*$, i.e. $\vec{U}$ changes scalar products into their complex conjugates. In particular, $\vec{U}$ preserves norms, since these are real. If Ω is a linear or antilinear operator on kets and is Hermitian, then so is the linear or antilinear operator $\bar{\Omega}$ defined by $\bar{\Omega} = \vec{U}\Omega\vec{U}^{\dagger}$. A similar result holds if Ω operates on bras. The inverses of the antilinear unitary operators $\vec{U}$ and $\overleftarrow{U}$ are also unitary. The product of a linear and an antilinear unitary operator is an antilinear unitary operator and the product of two antilinear unitary operators is a linear unitary operator. If Ω is an operator on kets (not necessarily linear or antilinear) and if $\bar{\Omega} = \vec{U}\Omega\vec{U}^\dagger$, then corresponding eigenvalues of Ω and $\bar{\Omega}$ are complex conjugates of each other, since $\Omega\,|S\rangle = c\,|S\rangle$ implies $\bar{\Omega}\,|\bar{S}\rangle = c^*\,|\bar{S}\rangle$, and conversely. If f is a function of operators $\Omega_1, \Omega_2, \ldots,$ on kets, then $\vec{U}f(\Omega_1, \Omega_2, \ldots)\vec{U}^\dagger = f^*(\bar{\Omega}_1, \bar{\Omega}_2, \ldots)$, where f^* denotes the complex conjugate function. This follows from the relations $\overline{(c\Omega)} = c^*\bar{\Omega}$, $\overline{(\Omega_1 + \Omega_2)} = \bar{\Omega}_1 + \bar{\Omega}_2$ and $\overline{(\Omega_1\Omega_2)} = \bar{\Omega}_1\bar{\Omega}_2$. A similar theorem is true if f is a function of operators $\Omega_1, \Omega_2, \ldots$ on bras.

PROBLEMS

(1) Show that if Ω_1 and Ω_2 are either both linear or both antilinear operators (on kets or bras) such that $\langle S|\,\Omega_1\,|S\rangle = \langle S|\,\Omega_2\,|S\rangle$ for all $|S\rangle$, then $\Omega_1 = \Omega_2$.

(2) Let A denote the operator that changes any wavefunction (in Schrödinger's representation for the particle space P) into its complex conjugate, so that $A\psi(q)\rangle\rangle = \psi^*(q)\rangle\rangle$. Show that A is antilinear, Hermitian and unitary, and determine its eigenvalue spectrum.

5.3 Continuous symmetries

TIME TRANSLATION

Let O and Ō be two observers differing only in that Ō's clock is retarded by the fixed amount τ relative to O's. The relation between the state vectors used by the two observers is then

$$|\bar{S}(\bar{t})\rangle = \mathcal{I}(\mathrm{O} \to \bar{\mathrm{O}})\,|S(t)\rangle \tag{5.8}$$

where $\bar{t} = t - \tau$ and the unitary operator U is called $\mathcal{I}$ for this transformation. $\mathcal{I}$ must be linear, since it is continuously connected to the identity. Indeed, $\mathcal{I}$ *is* the identity (apart from a possible phase factor), as we shall see. This is not surprising, since both observers make measurements on the system at the same objective time (labelled t by O and $\bar{t}$ by Ō) using the same measuring apparatus, the same reference frame, the same charge convention and the same "arrow" of time. They therefore use the same ray in Hilbert space for any given physical state, but label it differently as it develops in time. The classical dynamical variables associated with the two observers are such that

$$\bar{\boldsymbol{q}}_\alpha(\bar{t}) = \boldsymbol{q}_\alpha(t) \qquad \bar{\boldsymbol{p}}_\alpha(\bar{t}) = \boldsymbol{p}_\alpha(t) \tag{5.9}$$

$$\bar{\boldsymbol{e}}^\perp(\boldsymbol{x}, \bar{t}) = \boldsymbol{e}^\perp(\boldsymbol{x}, t) \qquad \bar{\boldsymbol{b}}(\boldsymbol{x}, \bar{t}) = \boldsymbol{b}(\boldsymbol{x}, t) \tag{5.10}$$

For the quantum mechanical expectation values we therefore have, by the correspondence principle,

$$\langle \bar{S}(\bar{t})| \, \boldsymbol{q}_\alpha \, |\bar{S}(\bar{t})\rangle = \langle S(t)| \, \boldsymbol{q}_\alpha \, |S(t)\rangle \tag{5.11}$$

with an analogous relation for $\boldsymbol{p}_\alpha$ and

$$\langle \bar{S}(\bar{t})| \, \boldsymbol{e}^\perp(\boldsymbol{x}) \, |\bar{S}(\bar{t})\rangle = \langle S(t)| \, \boldsymbol{e}^\perp(\boldsymbol{x}) \, |S(t)\rangle \tag{5.12}$$

with an analogous relation for $\boldsymbol{b}(\boldsymbol{x})$. Equation (5.8) then gives, since $|S(t)\rangle$ is arbitrary,

$$\mathscr{I}^\dagger \boldsymbol{q}_\alpha \mathscr{I} = \boldsymbol{q}_\alpha \qquad \mathscr{I}^\dagger \boldsymbol{p}_\alpha \mathscr{I} = \boldsymbol{p}_\alpha \tag{5.13}$$

$$\mathscr{I}^\dagger \boldsymbol{e}^\perp(\boldsymbol{x}) \mathscr{I} = \boldsymbol{e}^\perp(\boldsymbol{x}) \qquad \mathscr{I}^\dagger \boldsymbol{b}(\boldsymbol{x}) \mathscr{I} = \boldsymbol{b}(\boldsymbol{x}) \tag{5.14}$$

Since these equations are satisfied if $\mathscr{I}$ is the identity operator and since the set of operators formed by q, p, $\boldsymbol{e}^\perp$ and $\boldsymbol{b}$ is irreducible, $\mathscr{I}$ must be a phase factor times the identity. We may choose the arbitrary phase factor so that

$$|\bar{S}(\bar{t})\rangle = |S(t)\rangle \tag{5.15}$$

Now the Schrödinger equation used by O is

$$i\hbar \frac{\mathrm{d}}{\mathrm{d}t} |S(t)\rangle = H(t) \, |S(t)\rangle \tag{5.16}$$

and that used by Ō is

$$i\hbar \frac{\mathrm{d}}{\mathrm{d}\bar{t}} |\bar{S}(\bar{t})\rangle = \bar{H}(\bar{t}) \, |\bar{S}(\bar{t})\rangle \tag{5.17}$$

where we have considered the general case in which the Hamiltonian operators are possibly time-dependent. Using Equation (5.15) we obtain, since $|S(t)\rangle$ is arbitrary,

$$H(t) = \bar{H}(\bar{t}) = \bar{H}(t - \tau) \tag{5.18}$$

If the transformation O → Ō is to correspond to a symmetry of the system, then $\bar{H}(t) = H(t)$, i.e. the Hamiltonian operator used by O at his time t is the same as that used by Ō at *his* time t. It follows then from Equation (5.18) that $H(t+\tau) = \bar{H}(t) = H(t)$, and this is true for all τ if and only if H is independent of time. Thus time-translation invariance implies that the Hamiltonian is constant, or that energy is conserved, and conversely.

SPACE TRANSLATION

Suppose Ō's reference frame is translated, without change of orientation or handedness, through the constant displacement $\boldsymbol{d}$ relative to O's, and that the two observers agree on all other conventions. A physical point that has coordinate vector $\boldsymbol{x}$ for O then has coordinate vector $\bar{\boldsymbol{x}}$ for Ō, where $\bar{\boldsymbol{x}}=\boldsymbol{x}-\boldsymbol{d}$, and the relations between the classical dynamical variables are

$$\bar{\boldsymbol{q}}_\alpha(t)=\boldsymbol{q}_\alpha(t)-\boldsymbol{d} \qquad \bar{\boldsymbol{p}}_\alpha(t)=\boldsymbol{p}_\alpha(t) \tag{5.19}$$

$$\bar{\boldsymbol{e}}^\perp(\bar{\boldsymbol{x}},t)=\boldsymbol{e}^\perp(\boldsymbol{x},t) \qquad \bar{\boldsymbol{b}}(\bar{\boldsymbol{x}},t)=\boldsymbol{b}(\boldsymbol{x},t) \tag{5.20}$$

Denoting the quantum mechanical unitary displacement operator by $\mathscr{D}$, we have

$$|\bar{S}\rangle=\mathscr{D}(\mathrm{O}\to\bar{\mathrm{O}})\,|S\rangle \tag{5.21}$$

where the time label on the kets has been suppressed, since $\bar{t}=t$. As $\mathscr{D}$ is continuously connected to the identity, it must be a linear operator. It must also be such that

$$\mathscr{D}^\dagger\boldsymbol{q}_\alpha\mathscr{D}=\boldsymbol{q}_\alpha-\boldsymbol{d} \qquad \mathscr{D}^\dagger\boldsymbol{p}_\alpha\mathscr{D}=\boldsymbol{p}_\alpha \tag{5.22}$$

$$\mathscr{D}^\dagger\boldsymbol{e}^\perp(\boldsymbol{x})\mathscr{D}=\boldsymbol{e}^\perp(\boldsymbol{x}+\boldsymbol{d}) \qquad \mathscr{D}^\dagger\boldsymbol{b}(\boldsymbol{x})\mathscr{D}=\boldsymbol{b}(\boldsymbol{x}+\boldsymbol{d}) \tag{5.23}$$

in accordance with the correspondence principle. Now these relations are satisfied if we take for $\mathscr{D}$ the unitary operator

$$\mathscr{D}(\boldsymbol{d})=\exp\left(\frac{i}{\hbar}\boldsymbol{d}\cdot\boldsymbol{P}\right) \tag{5.24}$$

where $\boldsymbol{P}$ represents the total linear momentum of the system: $\boldsymbol{P}=\boldsymbol{P}_{\text{par}}+\boldsymbol{P}_{\text{rad}}$ with $\boldsymbol{P}_{\text{par}}=\sum_\alpha \boldsymbol{p}_\alpha$ and $\boldsymbol{P}_{\text{rad}}$ defined by Equation (4.102). We note first that, since $\boldsymbol{P}_{\text{rad}}$ commutes with particle variables,

$$\begin{aligned}\mathscr{D}^\dagger q_{\alpha i}\mathscr{D}&=\exp\left[-\frac{i}{\hbar}\boldsymbol{d}\cdot\boldsymbol{P}_{\text{par}}\right]q_{\alpha i}\exp\left[\frac{i}{\hbar}\boldsymbol{d}\cdot\boldsymbol{P}_{\text{par}}\right]\\&=q_{\alpha i}-d_i\end{aligned} \tag{5.25}$$

and

$$\begin{aligned}\mathscr{D}^\dagger p_{\alpha i}\mathscr{D}&=\exp\left[-\frac{i}{\hbar}\boldsymbol{d}\cdot\boldsymbol{P}_{\text{par}}\right]p_{\alpha i}\exp\left[\frac{i}{\hbar}\boldsymbol{d}\cdot\boldsymbol{P}_{\text{par}}\right]\\&=p_{\alpha i}\end{aligned} \tag{5.26}$$

where, for Equation (5.25), the canonical commutation relations and the operator identity of Theorem 1, Appendix D, have been used. Also, since

$\boldsymbol{P}_{\text{par}}$ commutes with field variables evaluated at field points $\boldsymbol{x}$ and since the commutation relation (4.88) holds,

$$\left[e_i^{\perp}(\boldsymbol{x}), \frac{i}{\hbar}\boldsymbol{d}\cdot\boldsymbol{P}\right] = (\boldsymbol{d}\cdot\nabla)e_i^{\perp}(\boldsymbol{x}) \tag{5.27}$$

and thus

$$\begin{aligned}\mathscr{D}^{\dagger}e_i^{\perp}(\boldsymbol{x})\mathscr{D} &= e_i^{\perp}(\boldsymbol{x}) + (\boldsymbol{d}\cdot\nabla)e_i^{\perp}(\boldsymbol{x}) + \frac{1}{2!}(\boldsymbol{d}\cdot\nabla)^2 e_i^{\perp}(\boldsymbol{x}) + \ldots \\ &= e_i^{\perp}(\boldsymbol{x}+\boldsymbol{d})\end{aligned} \tag{5.28}$$

In a similar way it may be shown that

$$\mathscr{D}^{\dagger}b_i(\boldsymbol{x})\mathscr{D} = b_i(\boldsymbol{x}+\boldsymbol{d}) \tag{5.29}$$

It follows that the unitary displacement operator is given to within a phase factor, which could vary with $\boldsymbol{d}$, by Equation (5.24). Thus the linear momentum operator is the generator of space translations.

If the Hamiltonian H is time-independent and if the space translation $\mathrm{O} \rightarrow \bar{\mathrm{O}}$ defines a symmetry of the system, then the unitary operator $\mathscr{D}$ commutes with H. If this holds for arbitrary displacement vectors $\boldsymbol{d}$, each component P_i of the generator of $\mathscr{D}$ must commute with H. Thus space-translation symmetry is equivalent to the law of conservation of linear momentum. We have already seen, in Section 4.3, that in the particular case in which there are no charged particles at all present the components of the linear momentum operator do commute with the Hamiltonian, since $[H_{\text{rad}}, \boldsymbol{P}_{\text{rad}}] = 0$. The same is true even when charged particles are present, however, so long as no external fields act on the system. It may readily be shown from the canonical commutation relations that the components of $\boldsymbol{P}_{\text{par}}$ commute with the particle Hamiltonian H_{par} in Equation (4.107); this reflects the translational invariance of the Coulomb interaction. Since particle variables commute with field variables evaluated at field points, it remains to show only that the components of the total linear momentum operator $\boldsymbol{P}$ commute with the interaction Hamiltonian H_{int} in Equation (4.108). This follows from the relations

$$\begin{aligned}[a_i(\boldsymbol{q}_\alpha), P_{\text{par}\,j}] &= i\hbar\frac{\partial a_i}{\partial q_{\alpha j}} \\ &= -[a_i(\boldsymbol{q}_\alpha), P_{\text{rad}\,j}]\end{aligned} \tag{5.30}$$

which follow in turn from the canonical commutation relations and the definitions of the momenta. Thus $\boldsymbol{P}$ commutes with the total Hamiltonian H, and the linear momentum is a constant of the motion.

SPACE ROTATION

If the observers O and $\bar{\mathrm{O}}$ use reference frames obtained one from the other by a proper rotation about a common origin, then the resolutes of a coordinate vector in the two frames are related by the equation $\bar{\boldsymbol{x}} = R\boldsymbol{x}$ in which the rotation matrix R is orthogonal and has determinant +1 (see Section 3.5). Assuming that the observers differ only through having their reference frames rotated with respect to each other, we may write for the transformation in Hilbert space

$$|\bar{S}\rangle = \mathscr{R}(\mathrm{O} \to \bar{\mathrm{O}})\,|S\rangle \tag{5.31}$$

where the unitary rotation operator $\mathscr{R}$ is linear and where the time label has again been suppressed. Since the classical dynamical variables all transform as vectors under proper rotations, we have

$$\bar{q}_{\alpha i}(t) = R_{ij}q_{\alpha j}(t) \qquad \bar{p}_{\alpha i}(t) = R_{ij}p_{\alpha j}(t) \tag{5.32}$$

$$\bar{e}_i^{\perp}(\bar{\boldsymbol{x}}, t) = R_{ij}e_j^{\perp}(\boldsymbol{x}, t) \qquad \bar{b}_i(\bar{\boldsymbol{x}}, t) = R_{ij}b_j(\boldsymbol{x}, t) \tag{5.33}$$

Here, as elsewhere, $\boldsymbol{x}$ and $\bar{\boldsymbol{x}}$ refer to the same physical field point. The quantum mechanical form of the transformations (5.32) and (5.33) is

$$\mathscr{R}^{\dagger}q_{\alpha i}\mathscr{R} = R_{ij}q_{\alpha j} \qquad \mathscr{R}^{\dagger}p_{\alpha i}\mathscr{R} = R_{ij}p_{\alpha j} \tag{5.34}$$

$$\mathscr{R}^{\dagger}e_i^{\perp}(\boldsymbol{x})\mathscr{R} = R_{ij}e_j^{\perp}(R^{\mathrm{T}}\boldsymbol{x}) \qquad \mathscr{R}^{\dagger}b_i(\boldsymbol{x})\mathscr{R} = R_{ij}b_j(R^{\mathrm{T}}\boldsymbol{x}) \tag{5.35}$$

as follows from the correspondence principle. For Equations (5.35), use has been made of the orthogonality of the matrix R in carrying out a change of variable. The relations (5.34) and (5.35) are satisfied if we take for $\mathscr{R}$ the unitary operator

$$\mathscr{R}(\boldsymbol{n}) = \exp\left(\frac{i}{\hbar}\,\boldsymbol{n}\,.\,\boldsymbol{J}\right) \tag{5.36}$$

where the pseudovector $\boldsymbol{n}$ gives, as in Section 3.5, the axis and angle of rotation and where $\boldsymbol{J}$ represents the total angular momentum of the system: $\boldsymbol{J} = \boldsymbol{J}_{\mathrm{par}} + \boldsymbol{J}_{\mathrm{rad}}$ with $\boldsymbol{J}_{\mathrm{par}} = \sum_{\alpha} \boldsymbol{q}_{\alpha} \times \boldsymbol{p}_{\alpha}$ and $\boldsymbol{J}_{\mathrm{rad}}$ defined by Equation (4.106). It is sufficient at first to consider only infinitesimal rotations specified by a vector $\delta\boldsymbol{n}$ and for which the rotation operator is given to first order in $\delta\boldsymbol{n}$ by

$$\mathscr{R}(\delta\boldsymbol{n}) = 1 + \frac{i}{\hbar}\,\delta\boldsymbol{n}\,.\,\boldsymbol{J} \tag{5.37}$$

For such rotations we obtain, since $\boldsymbol{J}_{\text{rad}}$ commutes with particle variables,

$$\mathcal{R}^{\dagger} q_{\alpha i} \mathcal{R} = q_{\alpha i} - \left[\frac{i}{\hbar}\, \delta\boldsymbol{n} \,.\, \boldsymbol{J}_{\text{par}},\, q_{\alpha i}\right]$$
$$= q_{\alpha i} - (\delta\boldsymbol{n} \times \boldsymbol{q}_{\alpha})_i$$
$$= R_{ij} q_{\alpha j} \tag{5.38}$$

again to first order in $\delta\boldsymbol{n}$. In a similar way the second of Equations (5.34) may be verified. For the transformation of the field operators we need the commutation relation (4.88) and the fact that $\boldsymbol{J}_{\text{par}}$ commutes with field variables taken at field points $\boldsymbol{x}$. (It should be noted also that the normally ordered form of $\boldsymbol{J}_{\text{rad}}$, or of $\boldsymbol{P}_{\text{rad}}$, is not essential for the evaluation of commutation relations.) It follows that, to first order in $\delta\boldsymbol{n}$,

$$\mathcal{R}^{\dagger} e_i^{\perp}(\boldsymbol{x}) \mathcal{R} = e_i^{\perp}(\boldsymbol{x}) - \left[\frac{i}{\hbar}\, \delta\boldsymbol{n} \,.\, \boldsymbol{J},\, e_i^{\perp}(\boldsymbol{x})\right]$$
$$= e_i^{\perp}(\boldsymbol{x}) + [(\delta\boldsymbol{n} \times \boldsymbol{x}) \,.\, \nabla] e_i^{\perp}(\boldsymbol{x}) - [\delta\boldsymbol{n} \times \boldsymbol{e}^{\perp}(\boldsymbol{x})]_i$$
$$= R_{ij} e_j^{\perp}(R^{\mathrm{T}}\boldsymbol{x}) \tag{5.39}$$

Here the transversality of $\boldsymbol{e}^{\perp}$ and the identity

$$\varepsilon_{ijk} A_l - \varepsilon_{jkl} A_i + \varepsilon_{kli} A_j - \varepsilon_{lij} A_k = 0 \tag{5.40}$$

which holds for any vector $\mathbf{A}$, have also been used. The second of Equations (5.35) may be verified in a similar way. Since we have specified its effect on an irreducible set of operators, the infinitesimal rotation operator must be given by Equation (5.37), apart from a possible phase factor that may depend on $\delta\boldsymbol{n}$. Thus the angular momentum $\boldsymbol{J}$ is the generator of space rotations. The operator for finite rotations is the limit of the product operator for a succession of small rotations:

$$\mathcal{R}(\boldsymbol{n}) = \lim_{r \to \infty} \left(1 + \frac{i}{\hbar} \frac{\boldsymbol{n} \,.\, \boldsymbol{J}}{r}\right)^r$$
$$= \exp\left(\frac{i}{\hbar}\, \boldsymbol{n} \,.\, \boldsymbol{J}\right) \tag{5.41}$$

Again $\mathcal{R}(\boldsymbol{n})$ could be multiplied by a physically irrelevant phase factor that may depend on $\boldsymbol{n}$.

The discussion of the connection between space-rotation symmetry and angular momentum conservation parallels that given for space-translation symmetry and linear momentum conservation—if H is time-independent and if the proper rotation $\mathrm{O} \to \bar{\mathrm{O}}$ defines a symmetry for arbitrary rotation vectors $\boldsymbol{n}$, then each component J_i of the generator of the rotation operator $\mathcal{R}$ commutes with H and is thus a constant of the motion. The law of conservation of angular momentum has been verified for the free field in Section 4.3, where it was shown that $[H_{\text{rad}}, \boldsymbol{J}_{\text{rad}}]=0$. To verify this law when the photons are coupled to charged particles but external fields are absent, we must show first that, as a consequence of the rotational invariance of the Coulomb interaction, $[H_{\text{par}}, \boldsymbol{J}_{\text{par}}]=0$. This may be done, as in ordinary quantum mechanics, by using the canonical commutation relations for the particle variables. Since $\boldsymbol{J}_{\text{par}}$ commutes with H_{rad} and $\boldsymbol{J}_{\text{rad}}$ with H_{par}, the total angular momentum $\boldsymbol{J}$ must then commute with the unperturbed Hamiltonian $H_{\text{par}}+H_{\text{rad}}$ or H_0. That it commutes also with the interaction Hamiltonian H_{int}, and hence with the total Hamiltonian H, may be seen by using the canonical commutation relations and the transverse nature of the vector potential $\boldsymbol{a}$ to show that

$$[a_i(\boldsymbol{q}_\alpha), J_j]=i\hbar\varepsilon_{ijk}a_k(\boldsymbol{q}_\alpha) \tag{5.42}$$

and

$$[p_{\alpha i}, J_j]=-i\hbar\varepsilon_{ijk}p_{\alpha k} \tag{5.43}$$

from which the required result follows.

It should be noted that for the coupled systems it is the total momentum (linear or angular) and not that of the particles or of the field separately that is a constant of the motion. The individual operators $\boldsymbol{P}_{\text{par}}$, $\boldsymbol{P}_{\text{rad}}$, $\boldsymbol{J}_{\text{par}}$, $\boldsymbol{J}_{\text{rad}}$ do not commute with the interaction Hamiltonian H_{int}, although they do commute with the unperturbed Hamiltonian H_0. Physically this means that momentum exchange takes place within the system between the particles and the field, but in such a way that the total momentum is conserved.

PROBLEM

Verify the commutation relations

$$[P_{\text{par}\,i}, P_{\text{par}\,j}]=0 \qquad [J_{\text{par}\,i}, P_{\text{par}\,j}]=i\hbar\varepsilon_{ijk}P_{\text{par}\,k},$$
$$[J_{\text{par}\,i}, J_{\text{par}\,j}]=i\hbar\varepsilon_{ijk}J_{\text{par}\,k}$$

Using these and the corresponding commutation relations for the field momentum operators, deduce that

$$[P_i, P_j]=0 \qquad [J_i, P_j]=i\hbar\varepsilon_{ijk}P_k$$
$$[J_i, J_j]=i\hbar\varepsilon_{ijk}J_k$$

5.4 Discrete symmetries

CHARGE CONJUGATION

Turning now to the discrete symmetries of the system, we consider first the effect of charge conjugation. If we adopt again the passive point of view, we have two observers O and Ō who use the same space-time coordinate frame but have differing conventions for the signs of the charges. Classically, the behaviour of the electric and magnetic induction fields under this transformation is given by

$$\bar{\boldsymbol{e}}^{\perp}(\boldsymbol{x}, t) = -\boldsymbol{e}^{\perp}(\boldsymbol{x}, t) \qquad \bar{\boldsymbol{b}}(\boldsymbol{x}, t) = -\boldsymbol{b}(\boldsymbol{x}, t) \tag{5.44}$$

i.e. the field components all change sign, while the particle variables $\boldsymbol{q}_\alpha$ and $\boldsymbol{p}_\alpha$ are unaltered (see Section 2.2). If the quantum mechanical unitary charge conjugation operator is denoted by $\mathscr{C}$, then

$$|\bar{S}\rangle = \mathscr{C}(\mathrm{O} \to \bar{\mathrm{O}})\,|S\rangle \tag{5.45}$$

where

$$\mathscr{C}^{\dagger}\boldsymbol{q}_\alpha\mathscr{C} = \boldsymbol{q}_\alpha \qquad \mathscr{C}^{\dagger}\boldsymbol{p}_\alpha\mathscr{C} = \boldsymbol{p}_\alpha \tag{5.46}$$

$$\mathscr{C}^{\dagger}\boldsymbol{e}^{\perp}(\boldsymbol{x})\mathscr{C} = -\boldsymbol{e}^{\perp}(\boldsymbol{x}) \qquad \mathscr{C}^{\dagger}\boldsymbol{b}(\boldsymbol{x})\mathscr{C} = -\boldsymbol{b}(\boldsymbol{x}) \tag{5.47}$$

That $\mathscr{C}$ is a linear operator may be seen by postmultiplying both sides of the commutation relation (4.88) by $\mathscr{C}$ and premultiplying by $\mathscr{C}^{\dagger}$. The left-hand side is unchanged, according to Equations (5.47), and the right-hand side is unchanged or changes sign, depending on whether $\mathscr{C}$ is linear or antilinear. (Note that $\mathscr{C}$ is, by Wigner's theorem, guaranteed to be either linear or antilinear.) Thus $\mathscr{C}$ must be linear. This is consistent also with the covariance of the canonical commutation relations. $\mathscr{C}$ is defined to within a phase factor by Equations (5.46) and (5.47), since its effect on an irreducible set of operators is specified. As $\mathscr{C}$ is linear, it can also be defined by its effect on a basis for the Hilbert space S, or by its effect on the product kets $|F\rangle\,|P\rangle$ where $|F\rangle$ is in the field space F and $|P\rangle$ is in the particle space P. Equations (5.46) are evidently satisfied if $\mathscr{C}$ is the extension of an operator, also to be called $\mathscr{C}$, on the space F alone, so that $\mathscr{C}\{|F\rangle\,|P\rangle\} = \{\mathscr{C}\,|F\rangle\}\,|P\rangle$. To determine the effect of $\mathscr{C}$ operating on the basis vectors in the plane wave representation for F, we note that the components of the vector potential $\boldsymbol{a}$ and its conjugate momentum $\boldsymbol{\pi}$, just like those of $\boldsymbol{e}^{\perp}$ and $\boldsymbol{b}$, change sign when similarity transformed by $\mathscr{C}$. The same is then true, since $\mathscr{C}$ is linear, for the Fourier transforms $\tilde{\boldsymbol{a}}$ and $\tilde{\boldsymbol{\pi}}$ of $\boldsymbol{a}$ and $\boldsymbol{\pi}$ and hence also for the operator $\boldsymbol{c}$ defined in Equation (4.36). Finally, it follows from Equation (4.43) that for any polarization basis and for any wave vector $\boldsymbol{k}$ the charge conjugation transformation of the photon annihilation operators is given by

$$\mathscr{C}^{\dagger}a^{(\lambda)}(\boldsymbol{k})\mathscr{C} = -a^{(\lambda)}(\boldsymbol{k}) = \mathscr{C}a^{(\lambda)}(\boldsymbol{k})\mathscr{C}^{\dagger} \tag{5.48}$$

where the second equation is a consequence of the unitarity of $\mathscr{C}$. A similar rule holds for the creation operators. Now for all $\boldsymbol{k}$, λ

$$\begin{aligned} a^{(\lambda)}(\boldsymbol{k})\mathscr{C}\,|\mathrm{vac}\rangle &= \mathscr{C}\mathscr{C}^{\dagger}a^{(\lambda)}(\boldsymbol{k})\mathscr{C}\,|\mathrm{vac}\rangle \\ &= -\mathscr{C}a^{(\lambda)}(\boldsymbol{k})\,|\mathrm{vac}\rangle = 0 \end{aligned} \tag{5.49}$$

Thus $\mathscr{C}\,|\text{vac}\rangle$, since it is normalized to unity, is of the form $e^{i\alpha}\,|\text{vac}\rangle$ where α is real. Because $\mathscr{C}$ is defined only to within an arbitrary multiplicative phase factor, we can take $e^{i\alpha}$ to be 1. This then fixes $\mathscr{C}$ completely. We thus have

$$\mathscr{C}\,|\text{vac}\rangle = |\text{vac}\rangle \tag{5.50}$$

and, for an n-photon state vector,

$$\begin{aligned}\mathscr{C}\,|\boldsymbol{k}_1, \lambda_1; \ldots; \boldsymbol{k}_n, \lambda_n\rangle \\ &= \mathscr{C}a^{(\lambda_1)\dagger}(\boldsymbol{k})\mathscr{C}^\dagger \ldots \mathscr{C}a^{(\lambda_n)\dagger}(\boldsymbol{k})\mathscr{C}^\dagger\mathscr{C}\,|\text{vac}\rangle \\ &= (-1)^n\,|\boldsymbol{k}_1, \lambda_1; \ldots; \boldsymbol{k}_n, \lambda_n\rangle\end{aligned} \tag{5.51}$$

Thus, for any type of polarization, $\mathscr{C}$ is diagonal in the plane wave representation, the eigenvalue, or charge-conjugation parity, of an n-photon state vector being $(-1)^n$. It follows immediately that $\mathscr{C}$ is Hermitian, $\mathscr{C} = \mathscr{C}^\dagger = \mathscr{C}^{-1}$, and $\mathscr{C}^2 = 1$. This is due to the choice of phase that makes the vacuum state vector be invariant under $\mathscr{C}$. The two observers O and $\bar{\text{O}}$ ascribe the same *ray* to any n-photon state, but corresponding vectors differ in sign if n is odd. The sign change is, of course, important in the superposition process, since different rays are ascribed to states that are not eigenstates of $\mathscr{C}$.

The charge conjugation operator $\mathscr{C}$ does not commute with the total Hamiltonian H and is not a constant of the motion. At first sight this seems surprising, since it was shown in Section 2.2 that the classical formulation of the Maxwell–Lorentz theory is covariant under charge conjugation. This transformation, however, is not induced by changes in the space-time reference system but by changes in the signs of the charges, which appear as parameters rather than as dynamical variables, and the transformation equations (5.44) must always be accompanied by the substitution $e_\alpha \to -e_\alpha$ (in, e.g., the expressions for the charge and current densities) on going from one observer to the other. Consequently, the covariance of the classical equations does not in this case require the invariance of the functional form of the Hamiltonian, as in Section 5.1. Each of the two observers O and $\bar{\text{O}}$ related by $\mathscr{C}$ can set up the classical canonical formalism and carry out canonical quantization. The operators representing the canonical dynamical variables have the same physical significance for O and $\bar{\text{O}}$, since both observers use the same space-time conventions. However, O, labelling the charges e_α, will arrive at the Hamiltonian H of Equation (4.8) while $\bar{\text{O}}$, labelling the charges $\bar{e}_\alpha$, will arrive at the Hamiltonian $\bar{H}$ obtained from H by replacing e_α by $\bar{e}_\alpha$, i.e. by $-e_\alpha$. Now, except in the case of the free field, H is not even in the charges and thus $\bar{H} \neq H$. (Classically, $\bar{H}$ and H are numerically equal but are different functions or functionals of the different sets of canonical variables $\boldsymbol{q}_\alpha$, $\boldsymbol{p}_\alpha$, $\boldsymbol{a}$, $\boldsymbol{\pi}$ and $\bar{\boldsymbol{q}}_\alpha$, $\bar{\boldsymbol{p}}_\alpha$, $\bar{\boldsymbol{a}}$, $\bar{\boldsymbol{\pi}}$. Quantum mechanically, H and $\bar{H}$ are different operators, being different functions or functionals of the set of operators $\boldsymbol{q}_\alpha$, $\boldsymbol{p}_\alpha$, $\boldsymbol{a}$, $\boldsymbol{\pi}$ which, according to the conventions of Section 5.1, are common to the two observers.) Using the phase convention that makes $\mathscr{C}^\dagger = \mathscr{C}$ we see that, since $\boldsymbol{q}_\alpha$ and $\boldsymbol{p}_\alpha$ are invariant under $\mathscr{C}$ and $\mathscr{C}\boldsymbol{a}\mathscr{C} = -\boldsymbol{a}$, $\mathscr{C}H\mathscr{C} = \bar{H}$. Thus $\mathscr{C}$, although an observable, does not commute with H and is not conserved. This implies that for the interacting system the number of photons does not have to change by an even number, as would otherwise be the case, since the charge-conjugation parity of an n-photon state is $(-1)^n$. (For the free field, on the other hand, the photon number N is itself a constant of the motion and $\mathscr{C}$ is obviously conserved.) It should be noted that since the Hamiltonians differ for the two observers, the velocities, as operators in Hilbert space, may also differ. Thus while $[\boldsymbol{a}, H] = [\boldsymbol{a}, \bar{H}]$ and hence $\dot{\boldsymbol{a}} = \dot{\bar{\boldsymbol{a}}}$, we have

$$\dot{\boldsymbol{q}}_\alpha = \frac{1}{i\hbar}[\boldsymbol{q}_\alpha, H] = \frac{1}{m_\alpha}\left\{\boldsymbol{p}_\alpha - \frac{e_\alpha}{c}\boldsymbol{a}(\boldsymbol{q}_\alpha)\right\} \tag{5.52a}$$

but

$$\dot{\bar{\boldsymbol{q}}}_\alpha = \frac{1}{i\hbar}[\boldsymbol{q}_\alpha, \bar{H}] = \frac{1}{m_\alpha}\left\{\boldsymbol{p}_\alpha - \frac{\bar{e}_\alpha}{c}\boldsymbol{a}(\boldsymbol{q}_\alpha)\right\} \tag{5.52b}$$

However, matrix elements taken between corresponding state vectors (referring to the same physical state, but for different observers) are the same. Thus if $|\bar{R}\rangle = \mathscr{C}|R\rangle$ and $|\bar{S}\rangle = \mathscr{C}|S\rangle$, then

$$\begin{aligned}\langle \bar{R}|\, \dot{\bar{\boldsymbol{q}}}_\alpha \,|\bar{S}\rangle &= \langle R|\, \mathscr{C}\dot{\bar{\boldsymbol{q}}}_\alpha\mathscr{C} \,|S\rangle \\ &= \langle R|\, \dot{\boldsymbol{q}}_\alpha \,|S\rangle \end{aligned} \tag{5.53}$$

This is in agreement with the correspondence principle, since classically the physical momentum $m\dot{\boldsymbol{q}}_\alpha$, like the canonical momentum $\boldsymbol{p}_\alpha$, is the same for both observers.

SPACE INVERSION

The space inversion or parity transformation relates two observers O and Ō who differ only in that their spatial axis systems are inverted with respect to each other, so that the coordinate transformation $\bar{\boldsymbol{x}} = -\boldsymbol{x}$ holds. The classical dynamical variables transform according to

$$\bar{\boldsymbol{q}}_\alpha(t) = -\boldsymbol{q}_\alpha(t) \qquad \bar{\boldsymbol{p}}_\alpha(t) = -\boldsymbol{p}_\alpha(t) \tag{5.54}$$

$$\bar{\boldsymbol{e}}^\perp(\bar{\boldsymbol{x}}, t) = -\boldsymbol{e}^\perp(\boldsymbol{x}, t) \qquad \bar{\boldsymbol{b}}(\bar{\boldsymbol{x}}, t) = \boldsymbol{b}(\boldsymbol{x}, t) \tag{5.55}$$

in keeping with the true vector character of $\boldsymbol{q}_\alpha$, $\boldsymbol{p}_\alpha$ and $\boldsymbol{e}$ and the pseudovector character of $\boldsymbol{b}$. If the unitary space inversion operator is denoted by $\mathscr{P}$, then the relation between the state vectors corresponding to a given physical state is

$$|\bar{S}\rangle = \mathscr{P}(\mathrm{O} \to \bar{\mathrm{O}})\,|S\rangle \tag{5.56}$$

and from this the transformation of the particle and field operators, namely

$$\mathscr{P}^\dagger \boldsymbol{q}_\alpha \mathscr{P} = -\boldsymbol{q}_\alpha \qquad \mathscr{P}^\dagger \boldsymbol{p}_\alpha \mathscr{P} = -\boldsymbol{p}_\alpha \tag{5.57}$$

$$\mathscr{P}^\dagger \boldsymbol{e}^\perp(\boldsymbol{x}) \mathscr{P} = -\boldsymbol{e}^\perp(-\boldsymbol{x}) \qquad \mathscr{P}^\dagger \boldsymbol{b}(\boldsymbol{x}) \mathscr{P} = \boldsymbol{b}(-\boldsymbol{x}) \tag{5.58}$$

follows by classical analogy. $\mathscr{P}$ is a linear operator, as may be checked by applying it to the canonical commutation relations, and is, as usual, defined to within a phase factor only. Equations (5.57) and (5.58) can be satisfied by taking $\mathscr{P} = \mathscr{P}_F \mathscr{P}_P$ where $\mathscr{P}_F$ is the extension of a unitary operator on the field space F and $\mathscr{P}_P$ is the extension of a unitary operator on the particle space P. If the operator $\mathscr{P}_P$ is defined in Schrödinger's representation by

$$\mathscr{P}_P \psi(q)\rangle\rangle = e^{i\alpha} \psi(-q)\rangle\rangle \tag{5.59}$$

where α is real and constant, then its Hermitian adjoint is such that

$$\mathscr{P}_P^\dagger \psi(q)\rangle\rangle = e^{-i\alpha} \psi(-q)\rangle\rangle \tag{5.60}$$

and so

$$\mathscr{P}_P^\dagger q_{\alpha i} \mathscr{P}_P \psi(q)\rangle\rangle = -q_{\alpha i} \psi(q)\rangle\rangle \tag{5.61}$$

and, from Equation (4.118),

$$\mathscr{P}_P^\dagger p_{\alpha i} \mathscr{P}_P \psi(q)\rangle\rangle = -p_{\alpha i} \psi(q)\rangle\rangle \tag{5.62}$$

Since ψ is an arbitrary wavefunction and $\mathscr{P}_F$ commutes with particle variables, Equations (5.57) will be satisfied by $\mathscr{P}$. The phase factor $e^{i\alpha}$ in Equation (5.59) may be chosen to be unity. Then $\mathscr{P}_P = \mathscr{P}_P^\dagger$ and $\mathscr{P}_P^2 = 1$, so that $\mathscr{P}_P$ is an observable with eigenvalues ± 1, the corresponding eigenvectors having wavefunctions that are either even or odd in q. Changing the phase of $\mathscr{P}_P$ or $\mathscr{P}_F$, or both, amounts merely to changing the phase of the product operator $\mathscr{P}$. The field operator $\mathscr{P}_F$ must be such that Equations (5.58) hold when $\boldsymbol{e}^\perp$ and $\boldsymbol{b}$ operate in the field space F alone. Since the vector potential and its conjugate momentum are true vectors, we have also

$$\mathscr{P}_F^\dagger \boldsymbol{a}(\boldsymbol{x}) \mathscr{P}_F = -\boldsymbol{a}(-\boldsymbol{x}) \qquad \mathscr{P}_F^\dagger \boldsymbol{\pi}(\boldsymbol{x}) \mathscr{P}_F = -\boldsymbol{\pi}(-\boldsymbol{x}) \tag{5.63}$$

and thus, for the operators $\mathbf{c}$ defined in Equation (4.36),

$$\mathscr{P}_F^\dagger \mathbf{c}(\boldsymbol{k})\mathscr{P}_F = -\mathbf{c}(-\boldsymbol{k}) \tag{5.64}$$

The transformation of the creation and annihilation operators depends on the relation between the polarization basis for wave vector $\boldsymbol{k}$ and that for wave vector $-\boldsymbol{k}$. In the case of linearly polarized photons it is possible to choose the polarization vector so that

$$\hat{\boldsymbol{e}}^{(\lambda)}(-\boldsymbol{k}) = -\hat{\boldsymbol{e}}^{(\lambda)}(\boldsymbol{k}) \tag{5.65}$$

for all $\boldsymbol{k}$, λ. We recall that the linear polarization vectors $\hat{\boldsymbol{e}}^{(\lambda)}(\boldsymbol{k})$ need not be real, although they must be essentially real, and that, even if they are real, the triad $\hat{\boldsymbol{e}}^{(1)}$, $\hat{\boldsymbol{e}}^{(2)}$ and $\hat{\boldsymbol{k}}$, or $\hat{\boldsymbol{e}}^{(3)}$, need not have the same handedness as the reference frame. With the convention adopted in Equation (5.65), the determinant of the unitary matrix $[\hat{e}_i^{(\lambda)}(\boldsymbol{k})]$ changes sign under the substitution $\boldsymbol{k} \rightarrow -\boldsymbol{k}$. Since the annihilation operators are the coefficients of the operators $\mathbf{c}(\boldsymbol{k})$ relative to the chosen polarization basis, it follows from Equations (5.64) and (5.65) that

$$\mathscr{P}_F^\dagger a^{(\lambda)}(\boldsymbol{k})\mathscr{P}_F = a^{(\lambda)}(-\boldsymbol{k}) \tag{5.66}$$

for all $\boldsymbol{k}$, λ and hence, since $\mathscr{P}_F$ is unitary, that

$$\mathscr{P}_F a^{(\lambda)}(\boldsymbol{k})\mathscr{P}_F^\dagger = a^{(\lambda)}(-\boldsymbol{k}) \tag{5.67}$$

The creation operators transform in a similar way. Now

$$\begin{aligned} a^{(\lambda)}(\boldsymbol{k})\mathscr{P}_F\,|\text{vac}\rangle &= \mathscr{P}_F a^{(\lambda)}(-\boldsymbol{k})\,|\text{vac}\rangle \\ &= 0 \end{aligned} \tag{5.68}$$

for all $\boldsymbol{k}$, λ and thus, since $\mathscr{P}_F$ is unitary, $\mathscr{P}_F\,|\text{vac}\rangle$ is a phase factor times $|\text{vac}\rangle$. $\mathscr{P}_F$ may be fixed completely by choosing the phase factor to be unity, so that

$$\mathscr{P}_F\,|\text{vac}\rangle = |\text{vac}\rangle \tag{5.69}$$

We then obtain for an n-photon state vector

$$\begin{aligned} \mathscr{P}_F\,|\boldsymbol{k}_1, \lambda_1; \ldots; \boldsymbol{k}_n, \lambda_n\rangle &= \mathscr{P}_F a^{(\lambda_1)\dagger}(\boldsymbol{k}_1)\mathscr{P}_F^\dagger \ldots \mathscr{P}_F a^{(\lambda_n)\dagger}(\boldsymbol{k}_n)\mathscr{P}_F^\dagger\mathscr{P}_F\,|\text{vac}\rangle \\ &= |-\boldsymbol{k}_1, \lambda_1; \ldots; -\boldsymbol{k}_n, \lambda_n\rangle \end{aligned} \tag{5.70}$$

Thus if the field contains n linearly polarized photons and these are described by observer O as having wave vectors $\boldsymbol{k}_1, \ldots, \boldsymbol{k}_n$ and polarization vectors $\hat{\boldsymbol{e}}^{(\lambda_1)}(\boldsymbol{k}_1), \ldots, \hat{\boldsymbol{e}}^{(\lambda_n)}(\boldsymbol{k}_n)$, they are described by observer $\bar{\text{O}}$ as having wave vectors $-\boldsymbol{k}_1, \ldots, -\boldsymbol{k}_n$ and polarization vectors $\hat{\boldsymbol{e}}^{(\lambda_1)}(-\boldsymbol{k}_1), \ldots, \hat{\boldsymbol{e}}^{(\lambda_n)}(-\boldsymbol{k}_n)$, the polarization vectors being such that Equations (5.65) hold. It follows immediately from Equations (5.69) and (5.70) and the fact that $\mathcal{P}_F$ is linear that $\mathcal{P}_F^2=1$ and hence that $\mathcal{P}_F=\mathcal{P}_F^\dagger$. Thus $\mathcal{P}_F$ is an observable with eigenvalues ± 1. $\mathcal{P}_F$ is not diagonal in the plane wave representation and so the n-photon states are states of mixed parity. The state vectors defined by

$$|\boldsymbol{k}_1, \lambda_1; \ldots; \boldsymbol{k}_n, \lambda_n\rangle^{\pm} = \frac{1}{\sqrt{2}}\{|\boldsymbol{k}_1, \lambda_1; \ldots; \boldsymbol{k}_n, \lambda_n\rangle \pm |-\boldsymbol{k}_1, \lambda_1; \ldots; -\boldsymbol{k}_n, \lambda_n\rangle\} \tag{5.71}$$

however, are eigenvectors of $\mathcal{P}_F$ with eigenvalues ± 1, respectively. There are many representations in which $\mathcal{P}_F$ is diagonal, as it has only two distinct eigenvalues; any linear combination of even parity states has even parity and any linear combination of odd parity states has odd parity. The effect of $\mathcal{P}_F$ on the state vectors for circularly polarized photons may also be determined. The discussion of Section 4.3 shows that the polarization vectors $\hat{\boldsymbol{e}}^{(\mathrm{L})}(\boldsymbol{k})$ and $\hat{\boldsymbol{e}}^{(\mathrm{R})}(-\boldsymbol{k})$ differ only by a phase factor. By suitably choosing this factor, it is possible to have

$$\hat{\boldsymbol{e}}^{(\mathrm{L})}(-\boldsymbol{k}) = -\hat{\boldsymbol{e}}^{(\mathrm{R})}(\boldsymbol{k}) \qquad \hat{\boldsymbol{e}}^{(\mathrm{R})}(-\boldsymbol{k}) = -\hat{\boldsymbol{e}}^{(\mathrm{L})}(\boldsymbol{k}) \tag{5.72}$$

as well as

$$\hat{\boldsymbol{e}}^{(\mathrm{L})*}(\boldsymbol{k}) = \hat{\boldsymbol{e}}^{(\mathrm{R})}(\boldsymbol{k}) \tag{5.73}$$

Equation (5.64) then gives

$$\mathcal{P}_F^\dagger a^{(\mathrm{L})}(\boldsymbol{k})\mathcal{P}_F = a^{(\mathrm{R})}(-\boldsymbol{k}) = \mathcal{P}_F a^{(\mathrm{L})}(\boldsymbol{k})\mathcal{P}_F^\dagger \tag{5.74}$$

$$\mathcal{P}_F^\dagger a^{(\mathrm{R})}(\boldsymbol{k})\mathcal{P}_F = a^{(\mathrm{L})}(-\boldsymbol{k}) = \mathcal{P}_F a^{(\mathrm{R})}(\boldsymbol{k})\mathcal{P}_F^\dagger \tag{5.75}$$

with similar relations for the creation operators. From these we obtain

$$\mathcal{P}_F\,|\boldsymbol{k}_1, \mathrm{L/R}; \ldots; \boldsymbol{k}_n, \mathrm{L/R}\rangle = |-\boldsymbol{k}_1, \mathrm{R/L}; \ldots; -\boldsymbol{k}_n, \mathrm{R/L}\rangle \tag{5.76}$$

so that $\mathcal{P}_F$ reverses all wave vectors and interchanges left and right. This is in agreement with the convention, adopted in Section 4.3, that a photon that is called left (right) circularly polarized in one frame is called right (left) circularly polarized in a frame of the opposite handedness.

By using the transformation properties (5.57) and (5.63) of the particle and field operators it is easy to verify that the total Hamiltonian H is invariant under, and thus commutes with, the product operator $\mathcal{P}$. Hence $\mathcal{P}$ (but not $\mathcal{P}_P$ or $\mathcal{P}_F$) is a constant of the motion for the coupled systems and, within the domain of pure electrodynamics, parity is a good quantum number.

TIME REVERSAL

The last of the symmetry transformations to be considered here, namely time reversal, is the only one that corresponds to an antilinear rather than a linear unitary operator in Hilbert space.† The classical time-reversal transformation is represented by the equations

$$\bar{\boldsymbol{q}}_\alpha(\bar{t}) = \boldsymbol{q}_\alpha(t) \qquad \bar{\boldsymbol{p}}_\alpha(\bar{t}) = -\boldsymbol{p}_\alpha(t) \tag{5.77}$$

$$\bar{\boldsymbol{e}}^\perp(\boldsymbol{x}, \bar{t}) = \boldsymbol{e}^\perp(\boldsymbol{x}, t) \qquad \bar{\boldsymbol{b}}(\boldsymbol{x}, \bar{t}) = -\boldsymbol{b}(\boldsymbol{x}, t) \tag{5.78}$$

where $\bar{t} = -t$. Thus positions and charge densities, and hence electric field components, are unaltered, while the components of momenta and current densities, and hence of magnetic induction fields, are reversed. In the quantum mechanical description we now have two observers O and $\bar{\text{O}}$ whose state vectors referring to the same physical system at the same objective time are related by

$$|\bar{S}(\bar{t})\rangle = \vec{\mathfrak{T}}(\text{O} \to \bar{\text{O}})\,|S(t)\rangle \tag{5.79}$$

where $\vec{\mathfrak{T}}$, the unitary time-reversal operator on kets, is such that

$$\vec{\mathfrak{T}}^\dagger \boldsymbol{q}_\alpha \vec{\mathfrak{T}} = \boldsymbol{q}_\alpha \qquad \vec{\mathfrak{T}}^\dagger \boldsymbol{p}_\alpha \vec{\mathfrak{T}} = -\boldsymbol{p}_\alpha \tag{5.80}$$

$$\vec{\mathfrak{T}}^\dagger \boldsymbol{e}^\perp(\boldsymbol{x}) \vec{\mathfrak{T}} = \boldsymbol{e}^\perp(\boldsymbol{x}) \qquad \vec{\mathfrak{T}}^\dagger \boldsymbol{b}(\boldsymbol{x}) \vec{\mathfrak{T}} = -\boldsymbol{b}(\boldsymbol{x}) \tag{5.81}$$

$\vec{\mathfrak{T}}$ must be an antilinear operator, if the canonical commutation relations are to be covariant, and is thus to be distinguished from the corresponding antilinear operator $\overleftarrow{\mathfrak{T}}$ on bras. To construct the operator $\vec{\mathfrak{T}}$ we first define an antilinear unitary operator $\vec{\mathfrak{T}}_P$ on the particle space P such that Equations (5.80) hold, and an antilinear unitary operator $\vec{\mathfrak{T}}_F$ on the field space F such that Equations (5.81) hold. It should be noted that the operator $\vec{\mathfrak{T}}$ on the product space S can not be factorized as $\vec{\mathfrak{T}}_F \vec{\mathfrak{T}}_P$ with $\vec{\mathfrak{T}}_F$ and $\vec{\mathfrak{T}}_P$ being extended antilinear operators on S, since the product of two antilinear operators is linear. (Such a factorization was possible for the charge conjugation and parity operators because these operators and their factors were all linear.) $\vec{\mathfrak{T}}_P$ may be defined as the operator that changes the wave function in Schrödinger's representation into its complex conjugate:

$$\vec{\mathfrak{T}}_P \psi(q)\rangle = \psi^*(q)\rangle \tag{5.82}$$

It may easily be verified that this operator is antilinear, Hermitian and unitary, and hence that $\vec{\mathfrak{T}}_P^2 = 1$. Changing Schrödinger's representation by an overall phase factor, which may be done, changes $\vec{\mathfrak{T}}_P$ by such a factor. Due to the antilinear character of $\vec{\mathfrak{T}}_P$, however, the relation $\vec{\mathfrak{T}}_P^2 = 1$ does not depend on any special choice of the phase. Thus $(\vec{\mathfrak{T}}_P e^{i\alpha})^2 = \vec{\mathfrak{T}}_P^2 = (e^{i\alpha}\vec{\mathfrak{T}}_P)^2$ for all real α. We must show that Equations (5.80) hold with $\vec{\mathfrak{T}}_P$ defined by Equation (5.82). First

$$\vec{\mathfrak{T}}_P q_{\alpha i} \vec{\mathfrak{T}}_P \psi(q)\rangle = q_{\alpha i} \psi(q)\rangle \tag{5.83}$$

† Of course, a product transformation that contains an odd number of time reversals also corresponds to an antilinear operator.

since the position operators are real functions of themselves. Also, from Equation (4.118),

$$\vec{\mathfrak{T}}_P p_{\alpha i} \vec{\mathfrak{T}}_P \psi(q)\rangle = -p_{\alpha i} \psi(q)\rangle \tag{5.84}$$

Thus, since ψ is an arbitrary wavefunction and $\vec{\mathfrak{T}}_P$ is Hermitian, Equations (5.80) are satisfied. Moreover, these equations determine $\vec{\mathfrak{T}}_P$ to within a phase factor, as p and q form an irreducible set of operators on P. Similarly the operator $\vec{\mathfrak{T}}_F$, which acts on the field space F alone, is determined to within a phase factor by Equations (5.81). From these equations we obtain, remembering that $\vec{\mathfrak{T}}_F$ is antilinear,

$$\vec{\mathfrak{T}}_F \boldsymbol{c}(\boldsymbol{k}) \vec{\mathfrak{T}}_F = -\boldsymbol{c}(-\boldsymbol{k}) \tag{5.85}$$

the operators $\boldsymbol{c}(\boldsymbol{k})$ being defined by Equations (4.36). If the linear polarization vectors $\hat{\boldsymbol{e}}^{(\lambda)}(\boldsymbol{k})$ are real (and not just essentially real) and are such that $\hat{\boldsymbol{e}}^{(\lambda)}(-\boldsymbol{k}) = -\hat{\boldsymbol{e}}^{(\lambda)}(\boldsymbol{k})$, as in the discussion of space inversion, then

$$\vec{\mathfrak{T}}_F^{\dagger} a^{(\lambda)}(\boldsymbol{k}) \vec{\mathfrak{T}}_F = a^{(\lambda)}(-\boldsymbol{k}) = \vec{\mathfrak{T}}_F a^{(\lambda)}(\boldsymbol{k}) \vec{\mathfrak{T}}_F^{\dagger} \tag{5.86}$$

since $\vec{\mathfrak{T}}_F$ is unitary also. Similarly the creation operators for $\boldsymbol{k}$ are transformed into those for $-\boldsymbol{k}$. From these properties we deduce that the ray representing the vacuum state is mapped into itself by $\vec{\mathfrak{T}}_F$ and choosing the phase of $\vec{\mathfrak{T}}_F$ so that the vacuum state vector is invariant we have

$$\vec{\mathfrak{T}}_F\,|\text{vac}\rangle = |\text{vac}\rangle \tag{5.87}$$

and hence, for an n-photon state with real polarization vectors

$$\vec{\mathfrak{T}}_F\,|\boldsymbol{k}_1, \lambda_1; \ldots ; \boldsymbol{k}_n, \lambda_n\rangle = |-\boldsymbol{k}_1, \lambda_1; \ldots ; -\boldsymbol{k}_n, \lambda_n\rangle \tag{5.88}$$

It follows that $\vec{\mathfrak{T}}_F$ is Hermitian, in the sense of antilinear operators, and that $\vec{\mathfrak{T}}_F^2 = 1$. This, however, is not due to the choice of phase that makes Equation (5.87) true. Comparing Equations (5.70) and (5.88) we see that $\mathcal{P}_F$ and $\vec{\mathfrak{T}}_F$ have the same matrix elements in the plane wave representation for which the polarization vectors are real and Equation (5.65) holds. These two operators are not equal, however, since one is linear and the other antilinear. This is evident also from their effect on circularly polarized state vectors. Keeping to the conventions of Equations (5.72) and (5.73) we have

$$\vec{\mathfrak{T}}_F a^{(\mathrm{L})}(\boldsymbol{k}) \vec{\mathfrak{T}}_F = a^{(\mathrm{L})}(-\boldsymbol{k}) \qquad \vec{\mathfrak{T}}_F a^{(\mathrm{R})}(\boldsymbol{k}) \vec{\mathfrak{T}}_F = a^{(\mathrm{R})}(-\boldsymbol{k}) \tag{5.89}$$

with corresponding equations for the creation operators. Thus

$$\vec{\mathfrak{T}}_F\,|\boldsymbol{k}_1, \mathrm{L/R}; \ldots ; \boldsymbol{k}_n, \mathrm{L/R}\rangle = |-\boldsymbol{k}_1, \mathrm{L/R}; \ldots ; -\boldsymbol{k}_n, \mathrm{L/R}\rangle \tag{5.90}$$

i.e. $\vec{\mathfrak{T}}_F$ reverses the wave vectors of circularly polarized states but, in contrast to $\mathcal{P}_F$, does not change the helicities.

An antilinear operator, just like a linear operator, is defined by its matrix elements in any representation, or by the images of all the vectors in a basis for the space on which the operator acts. (The relation between the representatives of a vector and those of its image and the transformation of matrix elements under change of basis are different in the two cases, however.) Let $\vec{\mathfrak{T}}$ be the antilinear operator on S such that

$$\vec{\mathfrak{T}}\,|\text{vac};\,\psi_m\rangle=\{\vec{\mathfrak{T}}_F\,|\,\text{vac}\}\{\vec{\mathfrak{T}}_P\,|\,\psi_m\}=|\text{vac};\,\psi_m^*\rangle \tag{5.91}$$

and

$$\begin{aligned}\vec{\mathfrak{T}}\,|\boldsymbol{k}_1,\lambda_1;\ldots;\boldsymbol{k}_n,\lambda_n;\psi_m\rangle&=\{\vec{\mathfrak{T}}_F\,|\boldsymbol{k}_1,\lambda_1;\ldots;\boldsymbol{k}_n,\lambda_n\rangle\}\{\vec{\mathfrak{T}}_P\,|\psi_m\rangle\}\\&=|-\boldsymbol{k}_1,\lambda_1;\ldots;-\boldsymbol{k}_n,\lambda_n;\psi_m^*\rangle\end{aligned} \tag{5.92}$$

where the polarization vectors are real and $\{\psi_m\}$ is a complete orthonormal set of wave functions for P. The effect of $\vec{\mathfrak{T}}$ acting on any ket of the product form $|F\rangle\,|P\rangle$ is then given by

$$\vec{\mathfrak{T}}\,|F;\,P\rangle=\{\vec{\mathfrak{T}}_F\,|F\rangle\}\{\vec{\mathfrak{T}}_P\,|P\rangle\} \tag{5.93}$$

since $(c_1c_2)^*=c_1^*c_2^*$ for any two complex numbers c_1, c_2. Thus, e.g., with circularly polarized photons and any wave function ψ,

$$\vec{\mathfrak{T}}\,|\boldsymbol{k}_1,\text{L/R};\ldots;\boldsymbol{k}_n,\text{L/R};\psi\rangle=|-\boldsymbol{k}_1,\text{L/R};\ldots;-\boldsymbol{k}_n,\text{L/R};\psi^*\rangle \tag{5.94}$$

It may be verified that $\vec{\mathfrak{T}}$ is Hermitian and unitary and hence that $\vec{\mathfrak{T}}^2=1$. Now from the definition of extended linear operators on the space S and the fact that $\vec{\mathfrak{T}}_P q_{\alpha i}\vec{\mathfrak{T}}_P$ and $q_{\alpha i}$ are equal, as linear operators on P, we obtain

$$\vec{\mathfrak{T}}\,q_{\alpha i}\vec{\mathfrak{T}}\,|F;\,P\rangle=q_{\alpha i}\,|F;\,P\rangle \tag{5.95}$$

and thus, since S is generated by kets $|F;\,P\rangle$, $\vec{\mathfrak{T}}\,q_{\alpha i}\vec{\mathfrak{T}}$ and $q_{\alpha i}$ are equal, as linear operators on S. Similarly the rest of Equations (5.80) and (5.81) are satisfied by $\vec{\mathfrak{T}}$. It follows directly from these equations that $\vec{\mathfrak{T}}$ commutes with the total Hamiltonian H. This does not yield a conservation law, however, since $\vec{\mathfrak{T}}$ is not linear and does not represent an observable quantity. Time-reversal symmetry is nevertheless important in furnishing relations between transition rates and cross-sections.

PROBLEMS

(1) Show that if the phase of the charge conjugation operator is chosen so that $\mathscr{C}\,|\text{vac}\rangle=-|\text{vac}\rangle$, then $\mathscr{C}$ is still Hermitian and such that $\mathscr{C}^2=1$, but that the charge conjugation parity of an n-photon state is now $(-1)^{n+1}$.

(2) Show that the condition $[H_{\text{par}}, \vec{\mathcal{T}}_P] = 0$ means that in Schrödinger's representation the particle Hamiltonian operator is a real function of q and $\partial/\partial q$. Deduce that the eigenvectors of H_{par} then correspond to wave functions that are either essentially real (i.e. are constant phase factors times real wave functions) or occur in complex conjugate pairs. Hence show that the eigenfunctions can always be chosen to be real.

REFERENCES

Dirac, P. A. M. (1958). "The Principles of Quantum Mechanics". Clarendon Press, Oxford.

Fonda, L. and Ghirardi, G. C. (1970). "Symmetry Principles in Quantum Physics". Marcel Dekker, New York.

Wigner, E. P. (1959). "Group Theory and its Application to the Quantum Mechanics of Atomic Spectra". Academic Press, New York and London.

Chapter 6

Interaction of Photons and Atoms

6.1 Approximations

THE FIXED-NUCLEI APPROXIMATION

In this chapter we turn to the problem of finding approximate solutions to the equations for the coupled systems of field and particles. It is convenient first, however, to make two other approximations of a different kind. We shall be interested in the interaction of radiation with bound aggregates (atoms, molecules, ions, etc.) which will often be referred to simply as atoms. It is sometimes a good approximation to treat the nuclei of the atoms as fixed and to regard the coordinates and momenta of the electrons only as dynamical variables. This is possible because of the large mass of the protons and neutrons compared with that of the electrons. The fixed-nuclei approximation involves, among other things, the neglect of the recoil of the atoms which should accompany the emission or absorption of photons. The recoil velocity is, however, normally very small. For example, the speed imparted to a hydrogen atom by an optical photon is of the order of 10^{-8} times the speed of light *in vacuo*. Such a speed would result in only a very slight Doppler shift of the frequency of the radiation emitted by the atom. In the fixed-nuclei approximation the electrons are regarded as moving in an external electrostatic field produced by stationary protons. The group of symmetries of the system that were discussed in the previous chapter is reduced by the presence of such a field. The effect of the external field on the electrons and, through the electrons, on the radiation field is allowed for, but the reaction of the dynamical system on the nuclei is assumed to be small and is neglected. This neglect results in the breakdown of some of the exact conservation laws, e.g. that of linear momentum. The symmetries that remain and their dynamical consequences are discussed in Section 6.2.

OCCUPATION-NUMBER STATES

The n-photon state vectors that were described in Chapter 4 are exact eigenvectors of H_{rad} and $\boldsymbol{P}_{\text{rad}}$ but are not normalizable and therefore do not represent physically realizable states. We now construct state vectors in the field space F that are only approximate eigenvectors of H_{rad} and $\boldsymbol{P}_{\text{rad}}$ but are normalized to unity. Let $\delta^3 k$ be a small region of $\boldsymbol{k}$-space surrounding the point $\boldsymbol{k}$ and let the operators $a_{\boldsymbol{k}\lambda}$ be given by

$$a_{\boldsymbol{k}\lambda} = \frac{1}{(\delta^3 k)^{1/2}} \iiint_{\delta^3 k} \mathrm{d}^3 k' a^{(\lambda)}(\boldsymbol{k}') \tag{6.1}$$

These operators and their Hermitian adjoints satisfy the commutation relations

$$[a_{\boldsymbol{k}\lambda}, a_{\boldsymbol{k}'\lambda'}] = 0 = [a^\dagger_{\boldsymbol{k}\lambda}, a^\dagger_{\boldsymbol{k}'\lambda'}] \tag{6.2}$$

as follows from the corresponding relations for the creation and annihilation operators $a^{(\lambda)\dagger}(\boldsymbol{k})$ and $a^{(\lambda)}(\boldsymbol{k})$. Moreover, if the regions $\delta^3 k$ and $\delta^3 k'$ do not overlap when $\boldsymbol{k} \neq \boldsymbol{k}'$, then

$$[a_{\boldsymbol{k}\lambda}, a^\dagger_{\boldsymbol{k}'\lambda'}] = \delta_{\lambda\lambda'}\, \delta_{\boldsymbol{k}\boldsymbol{k}'} \tag{6.3}$$

The occurrence of the Kronecker rather than the Dirac delta function on the right-hand side of this equation should be noted. Now the vector $|n_{\boldsymbol{k}\lambda}\rangle$ defined by

$$\begin{aligned} |n_{\boldsymbol{k}\lambda}\rangle &= \frac{1}{\sqrt{n!}} (a^\dagger_{\boldsymbol{k}\lambda})^n \,|\text{vac}\rangle \\ &= \frac{1}{\sqrt{n!}} \frac{1}{(\delta^3 k)^{n/2}} \iiint_{\delta^3 k} \mathrm{d}^3 k_1 \ldots \iiint_{\delta^3 k} \mathrm{d}^3 k_n \,|\boldsymbol{k}_1, \lambda; \ldots; \boldsymbol{k}_n, \lambda\rangle \end{aligned} \tag{6.4}$$

is normalized to unity. This may be seen by using the orthonormality properties (4.65). Similarly Equations (4.62) and (4.63) may be used to show that the effect of the "creation" and "annihilation" operators $a^\dagger_{\boldsymbol{k}\lambda}$ and $a_{\boldsymbol{k}\lambda}$ on the normalized state vectors is given by

$$a^\dagger_{\boldsymbol{k}\lambda} \,|n_{\boldsymbol{k}\lambda}\rangle = \sqrt{n+1}\, |(n+1)_{\boldsymbol{k}\lambda}\rangle \tag{6.5}$$

and

$$a_{\boldsymbol{k}\lambda} \,|n_{\boldsymbol{k}\lambda}\rangle = \sqrt{n}\, |(n-1)_{\boldsymbol{k}\lambda}\rangle \tag{6.6}$$

More general normalized vectors belonging to the n-photon subspace can also be constructed. Let $n^{(1)}, \ldots, n^{(l)}$ be a partition of n and put

$$|n^{(1)}_{\boldsymbol{k}_1\lambda_1}, \ldots, n^{(l)}_{\boldsymbol{k}_l\lambda_l}\rangle = \frac{1}{\sqrt{n^{(1)}!}} \cdots \frac{1}{\sqrt{n^{(l)}!}} (a^\dagger_{\boldsymbol{k}_1\lambda_1})^{n^{(1)}} \ldots (a^\dagger_{\boldsymbol{k}_l\lambda_l})^{n^{(l)}} \,|\text{vac}\rangle \tag{6.7}$$

where $(\boldsymbol{k}_1, \lambda_1), \ldots, (\boldsymbol{k}_l, \lambda_l)$ are distinct pairs and where, for $i, j = 1, \ldots, l$, the regions $\delta^3 k_i$ and $\delta^3 k_j$ do not overlap if $\boldsymbol{k}_i \neq \boldsymbol{k}_j$. The states corresponding to these vectors are known as occupation-number states, $n^{(i)}$ being the occupation number for mode $\boldsymbol{k}_i$, λ_i. The occupation-number state vectors are exact eigenvectors of the number operator N with eigenvalue n and are approximate eigenvectors of H_{rad} and $\boldsymbol{P}_{\text{rad}}$ with eigenvalues $\hbar c[n^{(1)}k_1 + \ldots + n^{(l)}k_l]$ and $\hbar[n^{(1)}\boldsymbol{k}_1 + \ldots + n^{(l)}\boldsymbol{k}_l]$, respectively. Two occupation-number state vectors are orthogonal unless they have identical occupation numbers, in which case the scalar product is unity. Since the creation and annihilation operators for different modes commute,

$$a^{\dagger}_{\boldsymbol{k}_i\lambda_i} |\ldots, n^{(i)}_{\boldsymbol{k}_i\lambda_i}, \ldots\rangle = \sqrt{n^{(i)}+1}\, |\ldots, [n^{(i)}+1]_{\boldsymbol{k}_i\lambda_i}, \ldots\rangle \tag{6.8}$$

and

$$a_{\boldsymbol{k}_i\lambda_i} |\ldots, n^{(i)}_{\boldsymbol{k}_i\lambda_i}, \ldots\rangle = \sqrt{n^{(i)}}\, |\ldots, [n^{(i)}-1]_{\boldsymbol{k}_i\lambda_i}, \ldots\rangle \tag{6.9}$$

where all occupation numbers other than $n^{(i)}$ remain unchanged. The above properties of the occupation number states may be deduced directly from the corresponding properties of the n-photon states discussed in Chapter 4 and from the definition (6.1). They may also be derived (Dirac 1958) from the harmonic-oscillator commutation relations (6.2) and (6.3) and from the fact that the vacuum state vector is normalized to unity and is annihilated by all the operators $a_{\boldsymbol{k}\lambda}$.

Let the whole of the three-dimensional $\boldsymbol{k}$-space be divided up into non-overlapping cells each of volume $(2\pi)^3/V$, where the volume V in ordinary space may be as large as is desired. (The factor $(2\pi)^3$ is inserted to conform to the usual mode expansions when the field is supposed enclosed in a large rectangular parallelepiped of volume V on the walls of which periodic boundary conditions are imposed. V is just the cube of the wavelength $\delta\lambda$ associated with the cell $\delta^3 k$.) Thus we now take $\delta^3 k = (2\pi)^3/V$ for a representative vector $\boldsymbol{k}$ from the interior of each cell, V being independent of $\boldsymbol{k}$. So long as V is large enough, the integral of a continuous function of $\boldsymbol{k}$ over a cell is approximately equal to the value of the integrand at the representative point for that cell multiplied by the cell volume. The integral over an extended region is then given approximately by the sum of the contributions from the individual cells into which the region is divided. Using the "cell" approximation we may expand the field operators in terms of the operators $a^{\dagger}_{\boldsymbol{k}\lambda}$ and $a_{\boldsymbol{k}\lambda}$. Thus

$$\boldsymbol{a}(\boldsymbol{x}) = \sum_{\boldsymbol{k}} \sum_{\lambda} \left(\frac{2\pi\hbar c}{Vk}\right)^{1/2} \{\hat{\boldsymbol{e}}^{(\lambda)}(\boldsymbol{k}) a_{\boldsymbol{k}\lambda} e^{i\boldsymbol{k}.\boldsymbol{x}} + \hat{\boldsymbol{e}}^{(\lambda)*}(\boldsymbol{k}) a^{\dagger}_{\boldsymbol{k}\lambda} e^{-i\boldsymbol{k}.\boldsymbol{x}}\} \tag{6.10}$$

$$\boldsymbol{e}^{\perp}(\boldsymbol{x}) = \sum_{\boldsymbol{k}} \sum_{\lambda} \left(\frac{2\pi\hbar ck}{V}\right)^{1/2} i\{\hat{\boldsymbol{e}}^{(\lambda)}(\boldsymbol{k}) a_{\boldsymbol{k}\lambda} e^{i\boldsymbol{k}.\boldsymbol{x}} - \hat{\boldsymbol{e}}^{(\lambda)*}(\boldsymbol{k}) a^{\dagger}_{\boldsymbol{k}\lambda} e^{-i\boldsymbol{k}.\boldsymbol{x}}\} \tag{6.11}$$

and

$$\boldsymbol{b}(\boldsymbol{x}) = \sum_{\boldsymbol{k}} \sum_{\lambda} \left(\frac{2\pi\hbar ck}{V}\right)^{1/2} i\{\hat{\boldsymbol{k}} \times \hat{\boldsymbol{e}}^{(\lambda)}(\boldsymbol{k}) a_{\boldsymbol{k}\lambda} e^{i\boldsymbol{k}.\boldsymbol{x}} - \hat{\boldsymbol{k}} \times \hat{\boldsymbol{e}}^{(\lambda)*}(\boldsymbol{k}) a^{\dagger}_{\boldsymbol{k}\lambda} e^{-i\boldsymbol{k}.\boldsymbol{x}}\} \tag{6.12}$$

where the summations are to extend over the representative vectors $\boldsymbol{k}$ of all the cells. Similarly the mode expansions (4.51), (4.101) and (4.102) now become

$$N = \sum_{\boldsymbol{k}} \sum_{\lambda} a^{\dagger}_{\boldsymbol{k}\lambda} a_{\boldsymbol{k}\lambda} \equiv \sum_{\boldsymbol{k}} \sum_{\lambda} N_{\boldsymbol{k}\lambda} \tag{6.13}$$

$$H_{\text{rad}} = \sum_{\boldsymbol{k}} \sum_{\lambda} \hbar ck N_{\boldsymbol{k}\lambda} \tag{6.14}$$

and

$$\boldsymbol{P}_{\text{rad}} = \sum_{\boldsymbol{k}} \sum_{\lambda} \hbar \boldsymbol{k} N_{\boldsymbol{k}\lambda} \tag{6.15}$$

respectively. To write the expansion (4.70) of a vector $|F\rangle$ in the field space F in terms of the occupation-number state vectors, we replace the integrals by sums as before and consider the contribution, for given n and given $\lambda_1, \ldots, \lambda_n$, from the neighbourhood of the point $(\boldsymbol{k}_1, \ldots, \boldsymbol{k}_n)$ in the $3n$-dimensional $\boldsymbol{k}$-space. Let $n^{(1)}$ of the pairs $(\boldsymbol{k}_i, \lambda_i)$, $i = 1, \ldots, n$, be alike of one kind, ..., and $n^{(l)}$ alike of another kind, so that $n^{(1)} + \ldots + n^{(l)} = n$. Because of the permutational invariance of the n-photon amplitudes, we can suppose without loss of generality that the first l pairs are distinct. In the expansion (4.70), for given n, the integration is over the whole $3n$-dimensional $\boldsymbol{k}$-space and each summation is over two orthogonal polarization directions. We therefore get identical contributions if we replace $\boldsymbol{k}_1, \lambda_1; \ldots; \boldsymbol{k}_n, \lambda_n$ by $\boldsymbol{k}_{p(1)}, \lambda_{p(1)}; \ldots; \boldsymbol{k}_{p(n)}, \lambda_{p(n)}$ where p is any permutation of $(1, \ldots, n)$. Since $n^{(1)}$ of the $(\boldsymbol{k}_i, \lambda_i)$ are alike of one kind, ..., and $n^{(l)}$ alike of another, the number of such contributions is $n!/\{n^{(1)}! \ldots n^{(l)}!\}$ and their sum is given by

$$\frac{1}{n^{(1)}! \ldots n^{(l)}!} \iiint_{\delta^3 k_1} \mathrm{d}^3 k'_1 \ldots \iiint_{\delta^3 k_n} \mathrm{d}^3 k'_n \, |\boldsymbol{k}'_1, \lambda_1; \ldots; \boldsymbol{k}'_n, \lambda_n\rangle \times \langle \boldsymbol{k}'_1, \lambda_1; \ldots; \boldsymbol{k}'_n, \lambda_n \mid F\rangle = |n^{(1)}_{\boldsymbol{k}_1\lambda_1}, \ldots, n^{(l)}_{\boldsymbol{k}_l\lambda_l}\rangle \langle n^{(1)}_{\boldsymbol{k}_1\lambda_1}, \ldots, n^{(l)}_{\boldsymbol{k}_l\lambda_l} \mid F\rangle \tag{6.16}$$

where the definition (6.7) and the approximation

$$\langle n^{(1)}_{\boldsymbol{k}_1\lambda_1}, \ldots, n^{(l)}_{\boldsymbol{k}_l\lambda_l} \mid F\rangle = \frac{1}{\sqrt{n^{(1)}!}} \cdots \frac{1}{\sqrt{n^{(l)}!}} (\delta^3 k)^{n/2} \langle \boldsymbol{k}_1, \lambda_1; \ldots; \boldsymbol{k}_n, \lambda_n \mid F\rangle \tag{6.17}$$

have both been used. Now let n denote the occupation number of mode $\boldsymbol{k}, \lambda$, instead of the total number of photons for a given state. If n is allowed to be zero as well as positive, a general occupation number state vector may be written as $|\ldots, n_{\boldsymbol{k}\lambda}, \ldots\rangle$ in which there is a countable infinity of ns, one for each of the discrete modes $\boldsymbol{k}, \lambda$. Only a finite number of the ns can be non-zero, the total number of photons being $\sum n$ with the (finite) sum extending over occupied modes. In this notation, the vacuum state vector is that for which all the ns are zero. We now have the orthonormality properties

$$\langle \ldots, n_{\boldsymbol{k}\lambda}, \ldots | \ldots, n'_{\boldsymbol{k}\lambda}, \ldots \rangle = \ldots \delta_{nn'} \ldots \tag{6.18}$$

where the infinite product on the right is 1 if all factors are 1 and zero if any factor is zero. We also have the expansion

$$|F\rangle = \sum |\ldots, n_{\boldsymbol{k}\lambda}, \ldots\rangle\langle \ldots, n_{\boldsymbol{k}\lambda}, \ldots | F\rangle \tag{6.19}$$

which follows from Equation (6.16) and is equivalent, within the cell approximation, to the expansion (4.70). The sum in Equation (6.19) extends over all occupation-number state vectors. There is one such state vector corresponding to each distinct way of distributing each partition of each non-negative integer over the modes of the field. If the vector $|F\rangle$ is normalized to unity, the orthonormality relations imply that

$$\sum |\langle \ldots, n_{\boldsymbol{k}\lambda}, \ldots | F\rangle|^2 = 1. \tag{6.20}$$

Thus $|\langle \ldots, n_{\boldsymbol{k}\lambda}, \ldots | F\rangle|^2$ is the probability (rather than the probability density) for there being $\ldots, n$ photons of mode $\boldsymbol{k}, \lambda, \ldots$.

6.2 Emission, absorption and scattering of radiation

The Hamiltonian for the system consisting of an atom and the radiation field may be written as

$$H = H_0 + H_{\text{int}} \equiv H_{\text{rad}} + H_{\text{atom}} + H_{\text{int}} \tag{6.21}$$

where H_{atom} includes (i) the kinetic energy of the electrons, (ii) the Coulomb repulsion between the electrons and (iii) the Coulomb attraction between the electrons and the nuclei, treated as fixed. The atom has in general discrete bound-state energy levels as well as a continuous spectrum corresponding to ionized states. The physically realizable states of the radiation field must be represented by normalized wave packets. We assume that if these packets are sufficiently sharply peaked near certain values of the wave vectors, then their detailed shape and phase relations are not physically significant. We can thus use for this purpose the occupation-number state vectors, which are sharply peaked wave packets of a particular kind. The eigenvectors, or approximate eigenvectors, of H_0 will be denoted by $|i\rangle, |f\rangle, \ldots$ and the corresponding eigenvalues by $\varepsilon_i, \varepsilon_f, \ldots$.

FIRST-ORDER PERTURBATION THEORY

Suppose that at time $t=0$ the system is in a state represented by a vector $|S(0)\rangle$ that is normalized to unity. To evaluate the probability $P(t)$ for an eigenstate (or approximate eigenstate) f of H_0 at some subsequent time t, we use the interaction or Dirac picture in which the state vector is related to that in the Schrödinger picture by a time-dependent unitary transformation generated by H_0:

$$|S'(t)\rangle = e^{iH_0t/\hbar}\,|S(t)\rangle \tag{6.22}$$

the two pictures being assumed to coincide at time $t=0$. The interaction-picture ket satisfies the differential equation

$$i\hbar\frac{\mathrm{d}}{\mathrm{d}t}|S'(t)\rangle = H'_{\text{int}}(t)\,|S'(t)\rangle \tag{6.23}$$

where $H'_{\text{int}}(t) = \exp(iH_0t/\hbar)H_{\text{int}}\exp(-iH_0t/\hbar)$. Thus the time evolution of $|S'(t)\rangle$ is governed by the unitarily transformed interaction Hamiltonian H'_{int} which, in contrast to H_{int}, is time-dependent, since H_0 and H_{int} do not commute.

The differential equation (6.23) subject to the given initial condition can be solved formally by putting

$$|S'(t)\rangle = T(t)\,|S(0)\rangle \tag{6.24}$$

where $T(t)$ is the time-evolution operator for the interaction picture and is given by

$$\begin{aligned} T(t) &= 1 + \frac{1}{i\hbar}\int_0^t \mathrm{d}t_1 H'_{\text{int}}(t_1) \\ &\quad + \frac{1}{(i\hbar)^2}\int_0^t \mathrm{d}t_1 H'_{\text{int}}(t_1)\int_0^{t_1}\mathrm{d}t_2 H'_{\text{int}}(t_2) + \ldots \\ &\equiv 1 + T_1(t) + T_2(t) + \ldots, \end{aligned} \tag{6.25}$$

as it then satisfies $T(0)=1$ as well as

$$i\hbar\frac{\mathrm{d}T}{\mathrm{d}t} = H'_{\text{int}}(t)T(t) \tag{6.26}$$

The probability for state f at time t may thus be expressed as

$$\begin{aligned} P(t) &= |\langle f \mid S(t)\rangle|^2 = |e^{-i\varepsilon_f t/\hbar}\langle f \mid S'(t)\rangle|^2 \\ &= |\langle f|\ T(t)\ |S(0)\rangle|^2 \end{aligned} \tag{6.27}$$

Note that this would not be true if f were not an eigenstate of H_0. Now the interaction Hamiltonian H_{int} consists of two parts, one proportional to e and the other to e^2, where $-e$ is the charge on the electron. The time-evolution operator can therefore be expanded as a power series in e, or, more properly, as a power series in $\sqrt{\alpha}$, where α is the dimensionless fine-structure constant: $\alpha = e^2/(\hbar c) \approx 1/137$. Since α, which is a measure of the strength of the coupling between the electromagnetic field and the charged particles, is small, the presence of the interaction term H_{int} in the total Hamiltonian H may be regarded as arising from a perturbation of the non-interacting systems of free field and isolated atom. The perturbation causes transitions between states that would otherwise be stationary. Writing $M_j(t)$ for the matrix element $\langle f|\ T_j(t)\ |S(0)\rangle$, we have as the exact expression for the probability

$$P(t) = |M_0(t) + M_1(t) + M_2(t) + \ldots|^2 \tag{6.28}$$

Usually we wish to calculate $P(t)$ for final states f that are orthogonal to the initial state $S(0)$, so that $M_0 \equiv \langle f \mid S(0)\rangle = 0$. If M_1 does not vanish, it may be either of order e or of order e^2, depending on which matrix elements of H_{int} are involved. If M_1 is of order e, then, since $M_2, M_3, \ldots$ are of order e^2 at most, the leading term in the expansion of $P(t)$ is $|M_1|^2$ and is of order e^2. First-order perturbation theory consists of retaining this term and neglecting those of higher order. We shall see by some examples in what circumstances this is a valid procedure.

SPONTANEOUS EMISSION

Let the initial state be such that (i) the atom is in a discrete excited state r with energy E_r and (ii) there are no photons present. Thus the initial state vector may be taken as

$$|S(0)\rangle = |\text{vac}; r\rangle = |i\rangle \tag{6.29}$$

say, and is normalized to unity. We then have, to order e^2,

$$\begin{aligned} P(t) &= \left| \frac{1}{i\hbar} \int_0^t \mathrm{d}t_1 \langle f|\ e^{i\varepsilon_f t_1/\hbar} H_{\text{int}} e^{-i\varepsilon_i t_1/\hbar}\ |i\rangle \right|^2 \\ &= 2\ |\langle f|\ H_{\text{int}}\ |i\rangle|^2 \frac{1 - \cos\{(\varepsilon_f - \varepsilon_i)t/\hbar\}}{(\varepsilon_f - \varepsilon_i)^2} \end{aligned} \tag{6.30}$$

We need to calculate the matrix element $\langle f|\, H_{\text{int}}\, |i\rangle$ only to order e and can therefore omit the e^2 term in H_{int}. Because the term of order e is linear in the vector potential and the vector potential is linear in the creation and annihilation operators, the only states f for which $\langle f|\, H_{\text{int}}\, |i\rangle$ does not vanish are one-photon states, and so $|f\rangle$ is of the form $|1_{\boldsymbol{k}\lambda}\,;\, s\rangle$ for some $\boldsymbol{k}$, λ and s. P is thus the probability that at time t a photon with wave vector $\boldsymbol{k}$ and polarization λ has been emitted and the atom has made a transition from state r to state s. This process, which is known as spontaneous emission, is represented graphically by a time-ordered Feynman diagram in Fig. 7. Remembering that $\varepsilon_i = E_r$ and $\varepsilon_f = \hbar ck + E_s$, we can now write for the emission probability

$$P(t) = 2\, |\langle s;\, 1_{\boldsymbol{k}\lambda}|\, H_{\text{int}}\, |\text{vac};\, r\rangle|^2 \frac{1-\cos\{(E-E_{rs})t/\hbar\}}{(E-E_{rs})^2} \tag{6.31}$$

where $E_{rs} = E_r - E_s$ and $E = \hbar ck$. For any value of t, $P(t)$ is very small unless $E \approx E_{rs}$. (This assumes that the squared modulus of the matrix element does not grow strongly with k or E.) Since, physically, E must be positive, there is a negligible probability of emission unless $E_{rs} > 0$, i.e. unless $E_r > E_s$. Thus the atom must make a transition from a higher to a lower energy level on emitting a photon and the *unperturbed* energy is approximately conserved in the process, since P is appreciable only when $E \approx E_{rs}$.

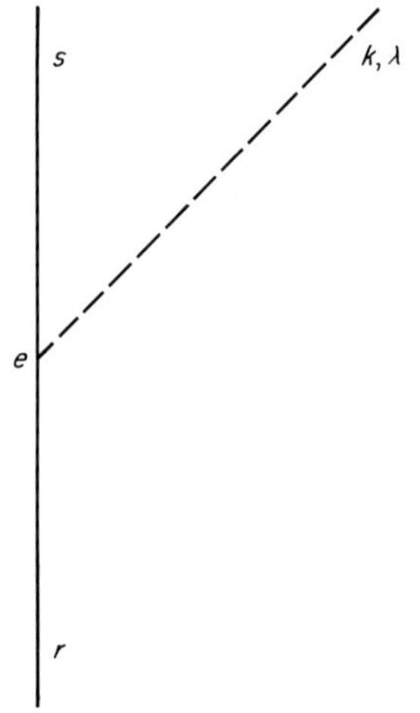

Fig. 7. Time-ordered Feynman diagram for spontaneous emission. The lower and upper portions of the solid line represent the initial and final atomic states r and s, respectively, and the dotted line represents the emitted photon of mode $\boldsymbol{k}$, λ. The vertex labelled e corresponds to the first-order term in the interaction Hamiltonian.

To proceed further, we note some properties of the function $F(u)$ defined by

$$F(u)=\frac{1-\cos u}{u^2} \tag{6.32}$$

This function has a main peak at the origin and secondary peaks that fall away with u like $2/u^2$. The area under the main peak is roughly π, since the width at the base is 4π and the height at the centre is $\frac{1}{2}$. The area under the whole curve from $-\infty$ to ∞ is exactly π, as may be shown by contour integration, and thus the bulk (about 90%) of the integral $\int_{-\infty}^{\infty} F(u)\,\mathrm{d}u$ comes from the area under the main peak. If u is the dimensionless variable $(E-E_{rs})t/\hbar$, so that $k=k_{rs}+u/(ct)$, then the probability at time t that the atom has made a transition from state r to state s and emitted a photon of polarization λ and with wave vector within d^3k of $\boldsymbol{k}$ is given by

$$P\,\mathrm{d}^3m=\frac{2t}{\hbar}|\langle s;\,1_{\boldsymbol{k}\lambda}|\,H_{\text{int}}\,|\text{vac};\,r\rangle|^2\,\rho(k)F(u)\,\mathrm{d}u \tag{6.33}$$

Here d^3m is the $\boldsymbol{k}$-space volume element divided by the cell volume $(2\pi)^3/V$ and is the number (assumed large) of representative $\boldsymbol{k}$-points in d^3k. The factor $\rho(k)$ is the density of states (number of states per unit energy interval) at $\boldsymbol{k}$:

$$\rho(k)=\frac{\mathrm{d}^3m}{\mathrm{d}E}=\frac{V}{(2\pi)^3}\frac{k^2\,\mathrm{d}\Omega}{\hbar c} \tag{6.34}$$

where $\mathrm{d}\Omega$ is an element of solid angle for the direction $\hat{\boldsymbol{k}}$, so that $\mathrm{d}^3k=k^2\,\mathrm{d}k\,\mathrm{d}\Omega$. The probability for the emission of a photon of any frequency (λ being fixed and $\hat{\boldsymbol{k}}$ being fixed within $\mathrm{d}\Omega$) is obtained by integrating the expression (6.33) with respect to u from $-k_{rs}ct$ to ∞, since k goes from 0 to ∞ and t is positive. If t is much greater than the period T_{rs} corresponding to the Bohr frequency for the transition $r\rightarrow s$, then in the region where F is relatively large we have $|k-k_{rs}|=|u|/(ct)\ll k_{rs}$ and it is a good approximation to replace k in the matrix element and in the expression for the density of states by k_{rs} and to replace the lower limit of the integral by $-\infty$. The probability is then proportional to t, the coefficient or transition probability per unit time being given, in accordance with Fermi's Golden Rule (Fermi 1950), by

$$a_{rs}^{(\lambda)}(\hat{\boldsymbol{k}})\,\mathrm{d}\Omega=\frac{2\pi}{\hbar}|\langle s;\,1_{\boldsymbol{k}_{rs}\lambda}|\,H_{\text{int}}\,|\text{vac};\,r\rangle|^2\,\rho(k_{rs}) \tag{6.35}$$

The energy of the emitted photon is distributed about the "energy-shell" value E_{rs} with a width inversely proportional to t, since $\Delta E=\hbar\,\Delta u/t\sim\hbar/t$.

For the formula (6.35) to be valid, the time t must be large compared to the atomic period T_{rs}. However, it must not be so large that the emission probability, which is linear in t, also becomes large. First-order perturbation theory, based on an expansion in powers of e without reference to the time dependence of the coefficients, ceases to be applicable when it predicts a transition probability that is not small compared to unity.

The calculation of the transition rate will be complete only when the matrix element of H_{int} between the initial and final state vectors has been evaluated. Using the expansion (6.10) of the vector potential and the properties (6.5) and (6.6) of the creation and annihilation operators, we obtain

$$\langle s; 1_{\boldsymbol{k}\lambda}|\, H_{\text{int}}\, |\text{vac}; r\rangle = \frac{e}{mc}\left(\frac{2\pi\hbar c}{Vk}\right)^{1/2} \hat{\boldsymbol{e}}^{(\lambda)*}(\boldsymbol{k}) \,.\, \langle s|\sum_{\alpha} \boldsymbol{p}_\alpha e^{-i\boldsymbol{k}.\boldsymbol{q}_\alpha}\,|r\rangle \qquad (6.36)$$

where m is the mass of the electron and the α sum is over electrons only. The *dipole approximation* can be used for optical or lower frequencies and bound states of atoms or small molecules, so that $\hbar ck \sim e^2/a_0$ and the "atomic" wave functions differ significantly from zero only for values of the electron coordinates within a distance of order a_0 from the centre of the atom. Then $ka_0 \sim e^2/(\hbar c) \approx 1/137 \ll 1$ and the exponentials that appear in the matrix element can be replaced by unity (for an atom centred at the origin). Now

$$\frac{e}{mc}\langle s|\sum_{\alpha} \boldsymbol{p}_\alpha\,|r\rangle = \frac{1}{i\hbar}\frac{e}{c}\langle s|\,[\sum_{\alpha} \boldsymbol{q}_\alpha, H_{\text{atom}}]\,|r\rangle$$
$$= ik_{rs}\,\langle s|\,\boldsymbol{\mu}\,|r\rangle \qquad (6.37)$$

where $\boldsymbol{\mu} = -e\sum_\alpha \boldsymbol{q}_\alpha$ and is the electric dipole moment operator. The matrix element can therefore be expressed in terms of the dipole transition moment $\langle s|\,\boldsymbol{\mu}\,|r\rangle$ or $\boldsymbol{\mu}^{sr}$ and the transition rate is (with $k = k_{rs}$)

$$a_{rs}^{(\lambda)}(\hat{\boldsymbol{k}})\,\mathrm{d}\Omega = \frac{1}{2\pi\hbar}\,k_{rs}^3\,|\hat{\boldsymbol{e}}^{(\lambda)*}(\boldsymbol{k}_{rs})\,.\,\boldsymbol{\mu}^{sr}|^2\,\mathrm{d}\Omega \qquad (6.38)$$

The total transition rate for any wave vector and polarization is obtained by integrating over Ω and summing over λ. Summing over the polarizations first gives, from Equation (4.41),

$$\sum_{\lambda} a_{rs}^{(\lambda)}(\hat{\boldsymbol{k}})\,\mathrm{d}\Omega = \frac{1}{2\pi\hbar}\,k_{rs}^3(\delta_{ij} - \hat{k}_i\hat{k}_j)\mu_i^{sr}\mu_j^{rs} \qquad (6.39)$$

where we have used the Hermiticity of $\boldsymbol{\mu}$ to get $(\boldsymbol{\mu}^{sr})^* = \boldsymbol{\mu}^{rs}$ and have put $\boldsymbol{k}_{rs} = k_{rs}\hat{\boldsymbol{k}}$, so that $\hat{\boldsymbol{k}}_{rs} = \hat{\boldsymbol{k}}$. Now

$$\oiint (\delta_{ij} - \hat{k}_i\hat{k}_j)\,\mathrm{d}\Omega = \frac{8\pi}{3}\,\delta_{ij} \qquad (6.40)$$

since the left-hand side must be a second-order isotropic tensor with trace 8π. Hence

$$\mathscr{A}^r_s \equiv \oiint \sum_{\lambda} a^{(\lambda)}_{rs}(\hat{\boldsymbol{k}})\,\mathrm{d}\Omega$$
$$= \frac{4}{3\hbar} k^3_{rs} |\boldsymbol{\mu}^{sr}|^2 \tag{6.41}$$

This is Einstein's A coefficient in dipole approximation. It is proportional to the cube of the transition frequency and the square of the length of the dipole transition moment. The reciprocal of $\mathscr{A}^r_s$ is the average lifetime of the upper state r with respect to the lower state s. For example, for optical transitions with a wavelength $\lambda \sim 5000$ Å and a dipole moment $\mu \sim ea_0$, the lifetime is of order 10^{-8} sec. Since the period is of order 10^{-15} sec, there is a range of values of t for which perturbation theory is valid. The detection of the emitted photons must take place at times t lying in this range, or else the emission rate will not be approximately constant.

ABSORPTION AND STIMULATED EMISSION

Suppose now the initial state vector is given by

$$|S(0)\rangle = |\ldots, n_{\boldsymbol{k}\lambda}, \ldots; s\rangle = |i\rangle \tag{6.42}$$

where the occupation number n is for each λ a slowly varying function of $\boldsymbol{k}$. The contribution $\mathrm{d}E$ to the field energy from all photons of a given polarization and with wave vector in d^3k is then $n\hbar ck$ times the number of cells in d^3k. The intensity per unit solid angle of the light is defined as a function of wave vector and polarization by

$$\mathscr{I}\,\mathrm{d}\Omega = \frac{\mathrm{d}E}{V\,\mathrm{d}\nu} = n\frac{\hbar k^3}{(2\pi)^2}\,\mathrm{d}\Omega \tag{6.43}$$

and is proportional to n but independent of V. The quantity $\mathrm{d}\nu$ is the frequency width corresponding to the width $\mathrm{d}k$, so that $\mathscr{I}$ has dimensions $\mathrm{erg\,cm^{-3}\,Hz^{-1}\,steradian^{-1}}$. The total intensity I ($\mathrm{erg\,cm^{-3}\,Hz^{-1}}$) for a given frequency is obtained by summing the expression (6.43) over the polarizations and integrating over the angles.

Radiation is absorbed from the incident beam if the final state f is represented by the state vector

$$|f\rangle = |\ldots, (n-1)_{\boldsymbol{k}\lambda}, \ldots; r\rangle \tag{6.44}$$

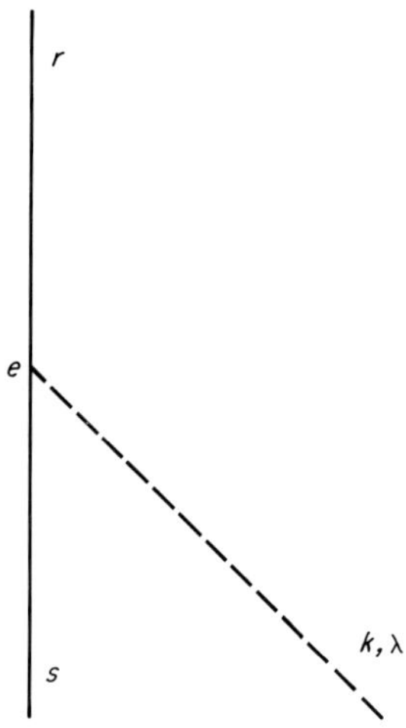

Fig. 8. Feynman diagram for absorption.

Thus the atom makes a transition $s \to r$ and the field is depleted by one photon of mode $\boldsymbol{k}$, λ. The difference between the initial and final eigenvalues depends only on the atomic energy levels and the energy of the missing photon, since $\varepsilon_i - \varepsilon_f = \hbar ck - E_r + E_s = E - E_{rs}$. For the probability of a particular final state to be appreciable requires that $E \approx E_{rs}$ with $E_r > E_s$; this means that the atom gains energy on absorbing a photon. The Feynman diagram depicting this process is shown in Fig. 8 for the case $n = 1$. The matrix element of the interaction Hamiltonian between the initial and final state vectors may be reduced as follows:

$$\begin{aligned}\langle f|\, H_{\text{int}}\, |i\rangle &= \langle r; \ldots, (n-1)_{\boldsymbol{k}\lambda}, \ldots|\, H_{\text{int}}\, |\ldots, n_{\boldsymbol{k}\lambda}, \ldots; s\rangle \\ &= \langle r; (n-1)_{\boldsymbol{k}\lambda}|\, H_{\text{int}}\, |n_{\boldsymbol{k}\lambda}; s\rangle \\ &= \sqrt{n}\,\langle r; \text{vac}|\, H_{\text{int}}\, |1_{\boldsymbol{k}\lambda}; s\rangle \end{aligned} \tag{6.45}$$

It is thus $\sqrt{n}$ times the complex conjugate of the matrix element for spontaneous emission. If the time t is much greater than the atomic period T_{rs}, we may make the same approximations as before to obtain a Golden Rule transition rate for the absorption of a photon of any frequency (λ being fixed and $\hat{\boldsymbol{k}}$ being fixed within $d\Omega$), namely

$$b_{sr}^{(\lambda)}(\hat{\boldsymbol{k}})\mathcal{I}\, d\Omega = \frac{2\pi}{\hbar} |\langle r; \text{vac}|\, H_{\text{int}}\, |1_{\boldsymbol{k}_{rs}\lambda}; s\rangle|^2\, n\rho(k_{rs}) \tag{6.46}$$

where n also is evaluated on the energy shell. Because of the factor n, the transition rate is proportional to the intensity of the incident light in the spectral region from which the photon is absorbed. In dipole approximation

$$b_{sr}^{(\lambda)}(\hat{\boldsymbol{k}})\mathcal{I}\,\mathrm{d}\Omega = \frac{2\pi}{\hbar^2}\,|\hat{\boldsymbol{e}}^{(\lambda)}(\boldsymbol{k}_{rs})\cdot\boldsymbol{\mu}^{rs}|^2\,\mathcal{I}\,\mathrm{d}\Omega \tag{6.47}$$

If the atom is bathed in isotropic unpolarized radiation, $\mathcal{I}$ is independent of $\hat{\boldsymbol{k}}$ and λ and the total absorption rate is given by

$$\oiint \sum_{\lambda} b_{sr}^{(\lambda)}(\hat{\boldsymbol{k}})\mathcal{I}\,\mathrm{d}\Omega = \frac{2\pi}{3\hbar^2}\,|\boldsymbol{\mu}^{rs}|^2\,I \equiv \mathcal{B}_r^s I \tag{6.48}$$

where $\mathcal{B}_r^s$ is Einstein's B coefficient for absorption and the total intensity I is just $8\pi\mathcal{I}$ evaluated at $k = k_{rs}$. The upper limit on the time for perturbation theory to be valid must now be much less than the reciprocal of $\mathcal{B}_r^s I$. Times less than this upper limit but much greater than the period T_{rs} can be found, provided the intensity I is not too large.

For an atom initially in the upper state r with radiation present as before, there is a probability for a transition to the lower state s accompanied by the emission of a photon with the same characteristics as some of those in the incident beam. The calculation of the transition rate for this process is very similar to that for the corresponding absorption $s \to r$, except that the square of the modulus of the matrix element now has a factor $n+1$ instead of n, due to the property (6.5) of the creation operators. The 1 represents the contribution of spontaneous emission and is present even when all the ns are zero. The emission induced or stimulated by the incident light is proportional to the intensity and, for isotropic unpolarized radiation, the rate is $\mathcal{B}_s^r I$ where the B coefficient for emission $r \to s$ is the same as that for absorption $s \to r$, so that $\mathcal{B}_s^r = \mathcal{B}_r^s$.

The levels E_r and E_s may be degenerate, with statistical weights g_r and g_s. The elements of complete orthogonal sets of substates for these levels will then be labelled by r' and s', respectively. The A and B coefficients for specified substates of the upper and lower levels are thus $\mathcal{A}_{s'}^{r'}$ and $\mathcal{B}_{s'}^{r'}$. The total spontaneous transition rate from a specified substate r' to any of the substates of s is $\sum_{s'} \mathcal{A}_{s'}^{r'}$, since the probabilities sum. For a statistical ensemble of identical systems, the average spontaneous transition rate A_s^r from an upper substate to any of the lower substates is

$$A_s^r = \frac{1}{g_r}\sum_{r'}\sum_{s'}\mathcal{A}_{s'}^{r'} \tag{6.49}$$

provided the substates of any degenerate level are equally populated. This is the total Einstein A coefficient for transitions $r \to s$, as used in Section 1.3. Similarly

$$B_s^r = \frac{1}{g_r}\sum_{r'}\sum_{s'}\mathcal{B}_{s'}^{r'} \tag{6.50}$$

and is the total Einstein B coefficient. Since $\mathscr{B}^{r'}_{s'} = \mathscr{B}^{s'}_{r'}$, Equation (1.11) immediately follows. Also Equation (1.12), which holds for the partial coefficients as well, may be recovered from Equations (6.41) and (6.48). The Einstein A and B coefficients derived from non-relativistic quantum electrodynamics (Dirac 1927a) therefore satisfy the universal relations obtained through a consideration of the conditions for thermal equilibrium between radiation and matter.

SECOND-ORDER PERTURBATION THEORY

A photon with wave vector $\boldsymbol{k}_1$ and polarization λ_1 has a probability of being scattered by an atom, initially in state r_1 with energy E_{r_1}, into a photon with wave vector $\boldsymbol{k}_2$ and polarization λ_2, the atom being left in state r_2 with energy E_{r_2}. The scattering is called coherent or Rayleigh scattering when the atom does not change its state, and is called Raman scattering when the atom gains or loses energy. We shall assume that the frequencies of the incident and scattered photons are not near the Bohr frequency for a transition of the atom from its initial or final state to any other state. (Otherwise the phenomenon of *resonance scattering* occurs.) We assume also that the scattered photon is not the same as the incident photon, which implies that the initial and final states are orthogonal.

As scattering involves a change of two photons, it is a second-order process and requires second-order perturbation theory for its description. Returning to the general expression (6.28) and retaining all matrix elements (M_1 and M_2) of order e^2 we obtain

$$P(t) = \left| \langle f|\, H_{\text{int}}\, |i\rangle \frac{e^{i(\varepsilon_f - \varepsilon_i)t/\hbar} - 1}{\varepsilon_f - \varepsilon_i} + \sum_v{}' \frac{\langle f|\, H_{\text{int}}\, |v\rangle\langle v|\, H_{\text{int}}\, |i\rangle}{\varepsilon_i - \varepsilon_v} \left\{ \frac{e^{i(\varepsilon_f - \varepsilon_i)t/\hbar} - 1}{\varepsilon_f - \varepsilon_i} - \frac{e^{i(\varepsilon_f - \varepsilon_v)t/\hbar} - 1}{\varepsilon_f - \varepsilon_v} \right\} \right|^2 \tag{6.51}$$

The prime on the summation symbol indicates that the sum is not over a complete set of states, since only those states v for which both $\langle f|\, H_{\text{int}}\, |v\rangle$ and $\langle v|\, H_{\text{int}}\, |i\rangle$ are of order e are to be included. (This precludes v from being either i or f.) The states v, which are known as virtual or intermediate states, must therefore differ from both i and f by one photon and thus contain either two photons (incident and scattered) or none at all. The direct matrix element $\langle f|\, H_{\text{int}}\, |i\rangle$ is now of order e^2, as it arises from the $\boldsymbol{a}^2$ term in H_{int}. The assumption that the scattering is off resonance means that none of the denominators $\varepsilon_i - \varepsilon_v$ or $\varepsilon_f - \varepsilon_v$ is near zero. On the other hand, the probability P is appreciable only when $\varepsilon_f - \varepsilon_i \approx 0$, in which case the second term in the curly brackets in Equation (6.51) is small compared to the first and can be neglected. We thus have an expression for $P(t)$ similar to that in Equation (6.30), the simple first-order matrix element being merely replaced by a compound second-order matrix element, namely

$$M_{fi}=\langle f|\, H_{\text{int}}\,|i\rangle+\sum_{v}{}' \frac{\langle f|\, H_{\text{int}}\,|v\rangle\langle v|\, H_{\text{int}}\,|i\rangle}{\varepsilon_i-\varepsilon_v} \tag{6.52}$$

If $u=\{E_2-(E_1-E_{r_2r_1})\}t/\hbar$ where $E_1=\hbar ck_1$ and $E_2=\hbar ck_2$ and if t is much greater than the period corresponding to $E_1-E_{r_2r_1}$, then the main peak of the function $F(u)$ is located in a region where k_2 is very near $k_1-k_{r_2r_1}$, and, if $k_1-k_{r_2r_1}>0$, there is a range of values of k_2 not containing $|k_{rr_1}|$ or $|k_{rr_2}|$ for any r but broad enough to make u go from a lower limit much less than -2π to an upper limit much greater than 2π. The transition rate for the scattered photon having k_2 within this range (λ_2 being fixed and $\hat{\boldsymbol{k}}_2$ being fixed within $\mathrm{d}\Omega$) is therefore

$$\mathrm{d}\Gamma_{fi}=\frac{2\pi}{\hbar}\,|M_{fi}|^2\,\rho(k_2) \tag{6.53}$$

the matrix element and density of states being evaluated on the energy shell, where $k_2+k_{r_2}=k_1+k_{r_1}$. The transition rate is, as will be seen, proportional to $1/V$. Thus if V is large enough (i.e. if the photon wave packets are sufficiently sharply peaked), the transition rate is small and there are times much greater than $2\pi/\{c(k_1-k_{r_2r_1})\}$ for which perturbation theory is valid. The differential cross-section (having dimensions cm^2) is defined by

$$\mathrm{d}\sigma_{fi}=\frac{\text{transition rate}}{\text{incident flux}}=\frac{\mathrm{d}\Gamma_{fi}}{(c/V)} \tag{6.54}$$

where the flux c/V corresponds to a particle density $1/V$ moving with speed c. The transition rate is proportional to the flux, so that the cross-section is independent of V. It should be noted that the transition rate and cross-section exist only for atomic states r_2 for which the energy-shell condition can be satisfied with k_2 positive.

KRAMERS–HEISENBERG SCATTERING

The Kramers–Heisenberg dispersion formula, which gives the cross-section for the scattering of light by an atom, was first derived from quantum electrodynamics by Dirac (1927b), the original derivation of Kramers and Heisenberg (1925) having been based on the semiclassical theory of radiation. To obtain this formula explicitly using Dirac's method, we must first evaluate the compound matrix element M_{fi}. The

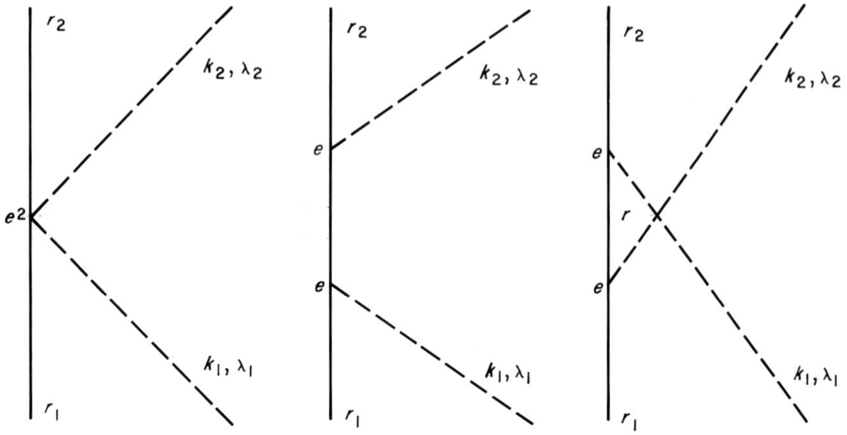

Fig. 9. Feynman diagrams for Kramers–Heisenberg scattering.

initial and final state vectors are $|1_{\boldsymbol{k}_1\lambda_1}; r_1\rangle$ and $|1_{\boldsymbol{k}_2\lambda_2}; r_2\rangle$, respectively, and the possible intermediate state vectors are $|\text{vac}; r\rangle$ and $|1_{\boldsymbol{k}_1\lambda_1}, 1_{\boldsymbol{k}_2\lambda_2}; r\rangle$ where r is any energy eigenstate of the atom. The Feynman diagrams corresponding to M_{fi} are shown in Fig. 9. It should be emphasized that the intermediate or virtual states of the system are merely suggested by the mathematical structure of the compound matrix element. Energy need not be even approximately conserved for virtual transitions, whereas it must be so for real transitions, which actually occur if the final states are observed. Using the Golden Rule formula (6.53) with the density of states (6.34) and the flux c/V, we obtain as the differential cross-section per unit solid angle

$$\frac{\mathrm{d}\sigma_{fi}}{\mathrm{d}\Omega}=\frac{k_2}{k_1}\Bigg|\hat{e}_i^{(\lambda_1)}(\boldsymbol{k}_1)\hat{e}_i^{(\lambda_2)*}(\boldsymbol{k}_2)\frac{e^2}{mc^2}\langle r_2|\sum_\alpha e^{i(\boldsymbol{k}_1-\boldsymbol{k}_2)\cdot\boldsymbol{q}_\alpha}|r_1\rangle$$
$$+\sum_r\Bigg[\frac{\hat{e}_j^{(\lambda_2)*}(\boldsymbol{k}_2)(e/mc)\langle r_2|\sum_\beta p_{\beta j}e^{-i\boldsymbol{k}_2\cdot\boldsymbol{q}_\beta}|r\rangle\,\hat{e}_i^{(\lambda_1)}(\boldsymbol{k}_1)(e/mc)\langle r|\sum_\alpha p_{\alpha i}e^{i\boldsymbol{k}_1\cdot\boldsymbol{q}_\alpha}|r_1\rangle}{E_1-E_{rr_1}}$$
$$-\frac{\hat{e}_i^{(\lambda_1)}(\boldsymbol{k}_1)(e/mc)\langle r_2|\sum_\alpha p_{\alpha i}e^{i\boldsymbol{k}_1\cdot\boldsymbol{q}_\alpha}|r\rangle\,\hat{e}_j^{(\lambda_2)*}(\boldsymbol{k}_2)(e/mc)\langle r|\sum_\alpha p_{\beta j}e^{-i\boldsymbol{k}_2\cdot\boldsymbol{q}_\beta}|r_1\rangle}{E_2+E_{rr_1}}\Bigg]\Bigg|^2 \tag{6.55}$$

As expected, this is independent of V. In dipole approximation all the exponential factors may be replaced by unity. It is then possible (Dirac

1927b, 1958) to recast the formula (6.55) so as to exhibit the dipole transition moments explicitly.† The first step in the transformation consists of substituting for the dipole-velocity in terms of the dipole-length matrix elements and gives, from Equation (6.37),

$$\frac{\mathrm{d}\sigma_{fi}}{\mathrm{d}\Omega}=\frac{k_2}{k_1}\left|\hat{e}_i^{(\lambda_1)}(\boldsymbol{k}_1)\hat{e}_j^{(\lambda_2)*}(\boldsymbol{k}_2)\right.$$
$$\left.\times\left\{M\frac{e^2}{mc^2}\delta_{ij}\delta_{r_1r_2}-\sum_r k_{r_2r}k_{rr_1}\left[\frac{\mu_j^{r_2r}\mu_i^{rr_1}}{E_1-E_{rr_1}}-\frac{\mu_i^{r_2r}\mu_j^{rr_1}}{E_2+E_{rr_1}}\right]\right\}\right|^2 \quad (6.56)$$

where M is the number of electrons in the atom. Then, since the components of the dipole operator $\boldsymbol{\mu}$ commute and the atomic eigenvectors are complete over the particle space P, we have, multiplying by $k_2/(\hbar c)$ for later convenience,

$$\frac{1}{\hbar c}\sum_r k_2(\mu_j^{r_2r}\mu_i^{rr_1}-\mu_i^{r_2r}\mu_j^{rr_1})=0 \quad (6.57)$$

It follows also from the canonical commutation relations that

$$\sum_r(\langle r_2|\,\mu_j\,|r\rangle\langle r|\sum_\beta p_{\beta i}\,|r_1\rangle-\langle r_2|\sum_\beta p_{\beta i}\,|r\rangle\langle r|\,\mu_j\,|r_1\rangle)=-i\hbar Me\,\delta_{ij}\,\delta_{r_1r_2} \quad (6.58)$$

or, from Equation (6.37),

$$\frac{1}{\hbar c}\sum_r(k_{rr_1}\mu_j^{r_2r}\mu_i^{rr_1}-k_{r_2r}\mu_i^{r_2r}\mu_j^{rr_1})=M\frac{e^2}{mc^2}\delta_{ij}\,\delta_{r_1r_1} \quad (6.59)$$

Adding Equations (6.57) and (6.59) and substituting for the direct scattering amplitude in Equation (6.56) shows that

$$\frac{\mathrm{d}\sigma_{fi}}{\mathrm{d}\Omega}=k_1k_2^3\left|\hat{e}_i^{(\lambda_1)}(\boldsymbol{k}_1)\hat{e}_j^{(\lambda_2)*}(\boldsymbol{k}_2)\sum_r\left(\frac{\mu_j^{r_2r}\mu_i^{rr_1}}{E_1-E_{rr_1}}-\frac{\mu_i^{r_2r}\mu_j^{rr_1}}{E_2+E_{rr_1}}\right)\right|^2 \quad (6.60)$$

In doing the analysis, the identities $k_2k_1+k_{r_2r}k_{rr_1}=(k_1-k_{rr_1})(k_2+k_{rr_1})$ and $k_2+k_{r_2r}=k_1-k_{rr_1}$, which stem from the energy-shell condition, have also been used. Equation (6.60) is the Kramers–Heisenberg dispersion formula in dipole approximation and in the form generally quoted.

INVARIANCE PROPERTIES

Invariance properties of the transition rates or cross-sections for scattering or other processes may be deduced from the symmetries of the system. These properties are most conveniently expressed in terms of the

† Dirac's transformation can be generalized to include all higher-order multipole moments in closed form (Healy 1977).

various initial and final states that may be used by a single observer. Let $|S(t_2)\rangle$ be the state vector at time t_2 that corresponds to the state vector $|S(t_1)\rangle$ or $|i\rangle$ at time t_1, the evolution being governed by the time-independent Hamiltonian H. The probability for the system being in the final state f at time t_2 is then

$$P_{f\leftarrow i}=|\langle f\,|\,S(t_2)\rangle|^2=|\langle f|\,e^{-iH(t_2-t_1)t/\hbar}\,|i\rangle|^2 \tag{6.61}$$

Suppose now that U is a unitary operator such that $[U, H]=0$ and let $|\bar{i}\rangle=U\,|i\rangle$ and $|\bar{f}\rangle=U\,|f\rangle$. If U is linear, it commutes not only with H but also with the time-evolution operator and so

$$\begin{aligned}P_{\bar{f}\leftarrow\bar{i}}&=|\langle\bar{f}|\,e^{-iH(t_2-t_1)/\hbar}\,|\bar{i}\rangle|^2=|\langle f|\,e^{-iH(t_2-t_1)/\hbar}\,|i\rangle|^2\\&=P_{f\leftarrow i}\end{aligned} \tag{6.62}$$

This shows the identity of the transition probabilities for the processes $i\rightarrow f$ and $\bar{i}\rightarrow\bar{f}$, which differ physically except when i and f are eigenstates of U. If U is antilinear, the time-evolution operator is similarity transformed into its inverse by U and we have instead of Equation (6.62),

$$\begin{aligned}P_{\bar{i}\leftarrow\bar{f}}&=|\langle\bar{i}|\,e^{-iH(t_2-t_1)/\hbar}\,|\bar{f}\rangle|^2=|\langle i|\,e^{iH(t_2-t_1)/\hbar}\,|f\rangle|^2\\&=|\langle f|\,e^{-iH(t_2-t_1)/\hbar}\,|i\rangle|^2\\&=P_{f\leftarrow i}\end{aligned} \tag{6.63}$$

This relation expresses the property of microscopic reversibility and is a consequence of time-reversal symmetry.

In the fixed-nuclei approximation and in the absence of external fields other than the electrostatic field of the nuclei, the possible symmetries of the system, besides those of time translation and charge conjugation, are (i) space rotation symmetry, (ii) space-inversion symmetry and (iii) time-reversal symmetry. (i) If the electrostatic potential is spherically symmetric about the origin, then an energy eigenfunction that undergoes a proper rotation about the origin is also an eigenfunction. If the wave vectors and polarization vectors of an incident and a scattered photon, for example, undergo the same rotation, the scattering cross-section is unaltered. We thus have the invariance property

$$\frac{d\sigma_{\bar{f}\bar{i}}}{d\Omega}=\frac{d\sigma_{fi}}{d\Omega} \tag{6.64}$$

which follows from Equation (6.62) and the fact that U commutes with H_{rad} and H_{atom} as well as with H, so that the new initial and final states have the same unperturbed energies as the old. In this case U is a unitary space-rotation operator. (Here we are using, in effect, the active rather than the passive interpretation of the unitary transformations, since the observer is fixed and the physical system is altered. Due to the assumed symmetries (central potential, etc.), however, only the electronic wave functions and not the nuclei need be transformed. The unitary operators are the inverses of those used with the passive interpretation. Thus a rotation of the coordinate axes with rotation vector $\boldsymbol{n}$ is described by the same operator as a rotation of the system with rotation vector $-\boldsymbol{n}$.) (ii) If the nuclear framework is invariant under space inversion, then so is the atomic or molecular Hamiltonian and the wave functions can be chosen to be either even or odd with respect to a change of sign of all the electronic coordinates. Equation (6.64) again holds, the new initial and final states being obtained, according to Section 5.4, by inverting the wave functions, reversing the wave vectors and, for circularly polarized photons, changing the helicities. (iii) If there is no external magnetic induction field present, the differential operator for the atomic or molecular Hamiltonian in Schrödinger's representation is real and the wave functions of the spin-zero "electrons" under consideration can be chosen to be real. Since the density of states depends quadratically on the magnitude k, we obtain from Equation (6.63), assuming that $d\Omega_1 = d\Omega_2 = d\Omega$,

$$k_2^2 \frac{d\sigma_{\bar{i}\bar{f}}}{d\Omega} = k_1^2 \frac{d\sigma_{fi}}{d\Omega} \tag{6.65}$$

This differs from Equation (6.64) not only through the factors k_1^2 and k_2^2 but also because the order of the transformed states on the left-hand side is changed. If the conventions of Section 5.4 are retained, the new states differ from the old by having the photon wave vectors reversed with the polarization vectors remaining unaltered.

PROBLEMS

(1) Assuming the light to propagate along the z-axis and to be circularly polarized, show that the atomic matrix element for spontaneous emission in the $2P \rightarrow 1S$ transition of a hydrogen atom is given exactly by

$$\frac{1}{\sqrt{2}} \langle 1, 0, 0|\, e^{-ikz}(p_x \pm ip_y)\, |2, 1, \mp 1\rangle = \mp \frac{i\hbar}{a_0} \frac{32\sqrt{2}}{(4k^2 a_0^2 + 9)^2}$$

where the upper and lower signs refer to left and right circular polarization, respectively, and where the atomic eigenvectors are labelled by the principal, orbital angular momentum and magnetic quantum numbers. Show also that the associated transition rate is $6 \times 10^8 \text{ sec}^{-1} \text{ steradian}^{-1}$.

(2) Show that if the conditions stated in the text are satisfied, the relations (6.64) and (6.65) are verified by the Kramers–Heisenberg dispersion formula in dipole approximation.

6.3 Line width and level shift

DAMPING THEORY

The elementary perturbation treatment of the previous section is not adequate for dealing with resonance phenomena or for calculating fine details such as the shift and shape of spectral lines. Heitler and co-workers (Heitler and Ma 1949, Arnous and Heitler 1953, Heitler 1954) have developed an elegant theory of radiative damping which will be described in this section in conjunction with the problem of the natural line shape in emission. This problem enables one of the pathological features of the quantum theory of radiation, namely the prediction of infinite electromagnetic level shifts, to be discussed. The by-passing of this difficulty through the device of mass renormalization will also be considered.

The formulae of damping theory may be succinctly expressed with the aid of Dirac's ζ-function (or distribution) defined (see Heitler 1954) by

$$\begin{aligned}\zeta(u) &= P\frac{1}{u} - i\pi\,\delta(u) = \lim_{\eta\to 0+}\frac{1}{u+i\eta}\\ &= -i\int_0^\infty e^{iu\tau}\,\mathrm{d}\tau \end{aligned} \tag{6.66}$$

where it is understood that all limiting processes (including integrating with respect to τ) are to be carried out only when ζ is multiplied by an analytic function and an integration with respect to u over the whole real line has been performed. The P denotes that the Cauchy principal value is to be taken. Since $u\,\delta(u)=0$, it follows immediately that $u\zeta(u)=1$. The properties

$$\lim_{t\to\pm\infty} -\frac{1}{2\pi i}\, e^{iut}\zeta(u) = \begin{cases} 0, & +\text{ sign}\\ \delta(u), & -\text{ sign}\end{cases} \tag{6.67}$$

may be established by using the integral representation in Equation (6.66). Moreover, if $\theta(t)$ is the unit step function, so that

$$\theta(t) = \begin{cases} 1, & t>0\\ 0, & t<0\end{cases} \tag{6.68}$$

then

$$i\zeta(u) = \int_{-\infty}^{\infty} e^{iut}\theta(t)\,\mathrm{d}t \tag{6.69}$$

and, by Fourier transformation,

$$\theta(t) = -\frac{1}{2\pi i}\int_{-\infty}^{\infty} e^{-iut}\zeta(u)\,\mathrm{d}u \tag{6.70}$$

The last relation may also be derived directly by contour integration.

Suppose the system is prepared in an eigenstate i of H_0 at time 0 and we wish to know the probability for a different eigenstate f at some time $t>0$. For the natural line shape problem we shall be interested in this probability for times much greater than the lifetime of the initial state. It is convenient mathematically to let the state vector be zero for $t<0$ and thus, to satisfy the initial condition, to have a discontinuity at $t=0$. The discontinuity is reflected in the differential equation for the time-evolution operator in the interaction picture, which is now

$$i\hbar\frac{\mathrm{d}T}{\mathrm{d}t} = H'_{\text{int}}(t)T(t) + i\hbar\,\delta(t) \tag{6.71}$$

instead of Equation (6.26). The solution satisfying $T(0+)=1$ is given by

$$\begin{aligned} T(t) &= \theta(t)e^{iH_0t/\hbar}e^{-iHt/\hbar} \\ &= -\frac{1}{2\pi i}\int_{-\infty}^{\infty} \mathrm{d}\varepsilon e^{-i(\varepsilon-H_0)t/\hbar}\zeta(\varepsilon-H) \end{aligned} \tag{6.72}$$

as may be verified from Equation (6.70) and the fact that $\theta'(t)=\delta(t)$. The θ-function ensures that the state vector is zero for $t<0$ and corresponds to i for $t=0+$. Of course, no physical implications will be drawn from the solution for $t<0$.

An arbitrary linear operator can be uniquely expressed as the sum of a diagonal operator and an off-diagonal operator defined through their matrix elements in a representation in which H_0 is diagonal. Let

$$\zeta(\varepsilon-H) = N(\varepsilon) + \zeta(\varepsilon-H_0)U(\varepsilon)N(\varepsilon) \tag{6.73}$$

where $N(\varepsilon)$ is the diagonal part. Since $\zeta(\varepsilon-H_0)$ is diagonal, $U(\varepsilon)$ is off-diagonal. Multiplying both sides of Equation (6.73) on the left by $\varepsilon-H$ and using $u\zeta(u)=1$ gives

$$\frac{1}{N(\varepsilon)} = U(\varepsilon) + \varepsilon - H_0 - \{H_{\text{int}} + H_{\text{int}}\zeta(\varepsilon-H_0)U(\varepsilon)\} \tag{6.74}$$

The operator on the left is diagonal, the element corresponding to the initial state i being $\varepsilon-\varepsilon_i+\frac{1}{2}i\hbar\Gamma(\varepsilon)$ where

$$\tfrac{1}{2}\hbar\Gamma(\varepsilon)=i\{H_{\text{int }ii}+\sum_{v}H_{\text{int }iv}\zeta(\varepsilon-\varepsilon_v)U_{vi}(\varepsilon)\} \tag{6.75}$$

Here $H_{\text{int }iv}=\langle i|\,H_{\text{int}}\,|v\rangle$, etc., and the sum is over a complete set of states. Also, since the off-diagonal elements must be zero, we have

$$U_{fi}(\varepsilon)=H_{\text{int }fi}+\sum_{v}H_{\text{int }fv}\zeta(\varepsilon-\varepsilon_v)U_{vi}(\varepsilon) \tag{6.76}$$

The expressions involving the ζ-functions are meaningful only in the limit as $V\to\infty$, when part of the v-sum becomes an integral over the photon wave vectors. Equation (6.76) is then an integral equation for the matrix elements of the unknown $U(\varepsilon)$, which is called the transition operator. Solution of this equation would enable $\Gamma(\varepsilon)$ to be determined from Equation (6.75). It will be seen that the real part of Γ represents a damping factor that prevents resonance catastrophes and that the imaginary part gives rise to level shifts. The usefulness of the formalism may be exhibited by calculating the transition probabilities in the limit as $t\to+\infty$. From the property (6.67) and Equations (6.72) to (6.76) we obtain for the probability amplitudes

$$\langle f|\,T(+\infty)\,|i\rangle=\frac{U_{fi}(\varepsilon_f)}{\varepsilon_f-\varepsilon_i+\frac{1}{2}i\hbar\Gamma(\varepsilon_f)} \tag{6.77}$$

The transition probability is thus the square of the modulus of the relevant transition matrix element, evaluated at $\varepsilon=\varepsilon_f$, times a typical resonance denominator. Γ also is evaluated at $\varepsilon=\varepsilon_f$. If the level shift represented by $\operatorname{Im}\Gamma$ is neglected, then the probability is appreciable only when $\varepsilon_f\approx\varepsilon_i$, so that the unperturbed energy is still approximately (to within the line width $\operatorname{Re}\Gamma$) conserved.

Although the expression (6.77) for the probability amplitude is exact, no exact solutions of the integral equation (6.76) are known. If, however, the matrix element of H_{int} between the initial and final state vectors is of order e, we may take this as a first approximation to the solution for U_{fi}, as the remaining terms on the right-hand side of Equation (6.76) are of higher order. Similarly in the expression (6.75) for Γ we may replace $U_{vi}(\varepsilon)$ by $H_{\text{int }vi}$ and include only states v that give a contribution of order e. (Note that the diagonal elements of H_{int} are of order e^2.) This method of solution does not consist of an expansion of the transition probability in powers of e^2, but recognizes instead the dominant influence of the resonance denominator. We can apply the method to find the spectral distribution of the radiation spontaneously emitted by an excited atom, taking $|i\rangle=|\text{vac}; r\rangle$ and $|f\rangle=|1_{\boldsymbol{k}\lambda}; s\rangle$ and assuming for simplicity that r is the first excited state and s the ground state. The amplitude for the probability that after a long time the atom has decayed to the ground state and a photon of mode $\boldsymbol{k}, \lambda$ has been emitted is

$$\langle s; 1_{\boldsymbol{k}\lambda}|\, T(+\infty)\,|\text{vac}; r\rangle=\frac{\langle s; 1_{\boldsymbol{k}\lambda}|\, H_{\text{int}}\,|\text{vac}; r\rangle}{E-E_{rs}-\delta E_r+\frac{1}{2}i\hbar\,\text{Re}\,\Gamma} \tag{6.78}$$

where

$$\text{Re}\,\Gamma(\varepsilon)=\frac{2\pi}{\hbar}\sum_{\boldsymbol{k}'}\sum_{\lambda'}|\langle s; 1_{\boldsymbol{k}'\lambda'}|\, H_{\text{int}}\,|\text{vac}; r\rangle|^2\,\delta(\varepsilon-E'-E_s) \tag{6.79}$$

In Equation (6.78) the damping factor $\text{Re}\,\Gamma$, which is positive, and the level shift δE_r, or $\frac{1}{2}\hbar\,\text{Im}\,\Gamma$, should, according to the exact expression (6.77), be evaluated at $\varepsilon=E+E_s$, and similarly the matrix element in the numerator should refer to the emitted photon with running energy E. If these quantities are slowly varying functions of E in the vicinity of the resonance, however, it is legitimate to treat them as constants by imposing the energy-shell condition $E=E_{rs}$. On the energy-shell $\text{Re}\,\Gamma$ is just the spontaneous emission rate A for the transition $r\rightarrow s$ and is given in dipole approximation by Equation (6.41). (The level shift δE_r will be calculated later.) The probability for the emission of a photon with energy within $\mathrm{d}E$ of E but with any polarization and propagation direction is then $P(E)\,\mathrm{d}E$ where

$$P(E)=\frac{\hbar A}{2\pi}\frac{1}{(E-E_{rs}-\delta E_r)^2+\frac{1}{4}\hbar^2A^2} \tag{6.80}$$

Thus, with the approximations indicated, the emitted line has a Lorentzian profile.

The connection between the damping factor and the spontaneous emission rate may be further understood by examining the amplitude of the initial state as a function of time. In the interaction picture the exact amplitude at time t is

$$\langle i|\, T(t)\,|i\rangle=-\frac{1}{2\pi i}\int_{-\infty}^{\infty}\mathrm{d}\varepsilon\,\frac{e^{-i(\varepsilon-\varepsilon_i)t/\hbar}}{\varepsilon-\varepsilon_i+\frac{1}{2}i\hbar\Gamma(\varepsilon)} \tag{6.81}$$

If the variation of Γ with ε is again neglected, the integration can be done explicitly for $t>0$ by closing the contour in the lower half-plane. The Schrödinger-picture amplitude, which differs from the expression (6.81) by a phase factor, is then given by

$$\begin{aligned}\langle i|\, e^{-iHt/\hbar}\,|i\rangle&=e^{-i\varepsilon_i t/\hbar}e^{-\Gamma t/2}\\&=e^{-i(E_r+\delta E_r)t/\hbar}e^{-At/2}\end{aligned} \tag{6.82}$$

Were it not for the damping factor, the initial state would be stationary with a shifted energy $E_r + \delta E_r$. Due to the spontaneous emission, however, the excited state gets depleted with the (approximate) probability e^{-At} following an exponential decay law. For times short compared to the lifetime of the initial state, $At \ll 1$ and the decay is approximately linear. The decay rate is just Einstein's A coefficient.

MASS RENORMALIZATION

If the energy-shell condition is used, so that the level shift is treated as constant, then the imaginary part of Equation (6.75) is, to order e^2,

$$\delta E_r = \langle r;\text{vac}|\, H_{\text{int}}\,|\text{vac}; r\rangle + P\sum_{\boldsymbol{k}'}\sum_{\lambda'}\sum_{r'}\frac{\langle r;\text{vac}|\, H_{\text{int}}\,|1_{\boldsymbol{k}'\lambda'}; r'\rangle\langle r'; 1_{\boldsymbol{k}'\lambda'}|\, H_{\text{int}}\,|\text{vac}; r\rangle}{E_{rr'} - \hbar ck'} \tag{6.83}$$

where the limit as $V \to \infty$ is assumed to be taken. The direct matrix element appearing here is independent of the state r (so long as the state vector is normalized to unity) and is zero if H_{int}, like H_{rad}, is required to be normally ordered with respect to the creation and annihilation operators. This term can thus be eliminated by subtracting an infinite c-number from the Hamiltonian, as was done for the zero-point energy. The structure of the compound matrix element is represented by the Feynman diagram of Fig. 10. It may be said to correspond to the emission and reabsorption of virtual photons by the atom. In principle, virtual photons of any energy E', however high, contribute to the sum. A systematic treatment of the level shift can thus be given only within the framework of a relativistic theory. An estimate of its magnitude may be obtained from the present theory, however, by imposing a high-energy cut-off $E_{\text{max}} < mc^2$ for virtual as well as for real photons. This is equivalent to assuming that the electron "sees" only radiation of wavelength greater than the Compton wavelength $\hbar/(mc)$ or $3{\cdot}862 \times 10^{-11}$ cm. (This is larger than the classical "electron radius" $e^2/(mc^2)$ or $2{\cdot}818 \times 10^{-13}$ cm by a factor of $1/\alpha$.) The expression (6.83) may then be evaluated in dipole approximation. Modelling the atom as a single electron in a static potential, we obtain

$$\delta E_r = -\frac{2\alpha}{3\pi}\sum_{r'}\frac{|\langle r'|\,\boldsymbol{p}\,|r\rangle|^2}{m^2c^2}\,P\int_0^{E_{\text{max}}}\frac{E'\,\mathrm{d}E'}{E' + E_{r'r}} \tag{6.84}$$

The integrals diverge linearly with the cut-off E_{max}. (Note, however, that the dipole approximation is not valid if E_{max} is very large.) Taking $E_{\text{max}} \sim mc^2$ and the root mean square speed of the electron in a bound

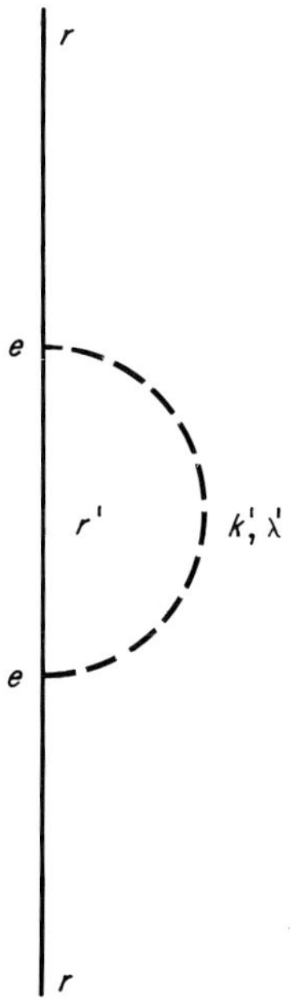

Fig. 10. Feynman diagram for the energy shift δE_r.

state $\sim \alpha c$ would give $\delta E_r \sim \alpha e^2/a_0$. This is much larger than the experimentally observed Lamb shift (Lamb and Retherford 1947) and is the same order of magnitude as the energy shift for a free electron. If the atomic eigenfunctions are replaced by normalized but sharply peaked free-electron wave packets, the expression (6.84) becomes

$$\delta E_{\mathrm{p}} = -\frac{2\alpha}{3\pi}\frac{p^2}{m^2c^2}\int_0^{E_{\max}} \mathrm{d}E' \tag{6.85}$$

where p is the (c-number) magnitude of the electron's momentum. This may be verified directly from Equation (6.83). (The electron is non-relativistic and the energy of the virtual photons less than mc^2.) The non-relativistic contribution $\delta\tilde{E}_r$ to the observed Lamb-shift energy is the difference between δE_r for a bound state and δE_{p}, where p is the root mean square value of the electron's momentum in the bound state:

$$\delta\tilde{E}_r = \delta E_r - \delta E_{\mathrm{p}} = \frac{2\alpha}{3\pi}\frac{1}{m^2c^2}\sum_{r'} E_{r'r}\,|\langle r'|\,\boldsymbol{p}\,|r\rangle|^2\, P\int_0^{E_{\max}} \frac{\mathrm{d}E}{E+E_{r'r}} \tag{6.86}$$

This diverges only logarithmically with the cut-off and, if $E_{\max} \sim mc^2$, gives a result in good agreement with experiment.

The electron self-energy is present whether the electron is free or bound and only the difference (6.86) shows up experimentally as a shift in the spectral line. The self-energy can be regarded as arising from an electromagnetic contribution to the mass of the electron and its effect can be eliminated by the process of mass renormalization. We suppose the electron to have a "bare" mechanical mass m_0. It is this mass, and not the experimentally observed mass m, that appears originally in the Hamiltonian H for the coupled systems. The bare mass m_0 or the bare energy $p^2/(2m_0)$ of a free electron with momentum p is unobservable, since the electron can never be separated from its own radiation field, i.e. its charge can never be switched off. Thus the electron is continually emitting and reabsorbing virtual photons. The total energy of a free electron with observed mass m is given by

$$\frac{p^2}{2m}=\frac{p^2}{2m_0}+\delta E_p \tag{6.87}$$

Putting $m=m_0+\delta m$ and assuming that the electromagnetic contribution δm is small compared to m_0 we get from Equation (6.85) and to order e^2

$$\delta m=\frac{4\alpha}{3\pi}\frac{1}{c^2}\int_0^{E_{\max}}\mathrm{d}E'\sim\frac{4\alpha}{3\pi}m \tag{6.88}$$

for $E_{\max}\sim mc^2$. With neglect of terms of order e^3, the Hamiltonian may be written as

$$\begin{aligned}H&=H_{\text{rad}}+\frac{1}{2m_0}\boldsymbol{p}^2+U(\boldsymbol{q})+\frac{e}{m_0c}\boldsymbol{p}\cdot\boldsymbol{a}(\boldsymbol{q})+\frac{e^2}{2m_0c^2}:\boldsymbol{a}^2(\boldsymbol{q}):\\&\approx H_{\text{rad}}+\frac{1}{2m}\boldsymbol{p}^2+U(\boldsymbol{q})+\frac{e}{mc}\boldsymbol{p}\cdot\boldsymbol{a}(\boldsymbol{q})+\frac{e^2}{2mc^2}:\boldsymbol{a}^2(\boldsymbol{q}):\\&\quad+\frac{\delta m}{m}\cdot\frac{1}{2m}\boldsymbol{p}^2\\&\equiv H_{\text{rad}}+H_{\text{atom}}+H_{\text{int}}\end{aligned} \tag{6.89}$$

where U is the potential energy function. We can identify $\boldsymbol{p}^2/(2m)+U$ with the previously used H_{atom}, so that the experimental mass m appears in the formulae for the energy levels, etc. H_{int}, however, now contains the "mass renormalization counter-term" $(\delta m/m)\boldsymbol{p}^2/(2m)$ which ensures that, to order e^2, (i) the energy of a free electron of momentum $\boldsymbol{p}$ is $\boldsymbol{p}^2/(2m)$ and (ii) the level shift computed directly from the expression (6.83) is just the Lamb-shift energy $\delta\tilde{E}_r$. It should be noted that the counter-term makes no contribution to any of the processes discussed in Section 6.2, so that the calculations given there remain valid, with m being the experimental mass of the electron. The non-relativistic Lamb-shift energy was first derived by the present method by Bethe (1947). (See also Schweber 1961, Messiah 1962, Power 1964, Kroll 1964).

If the mass renormalization programme is followed through, the Lamb-shift energy $\delta\tilde{E}_r$ will appear in place of δE_r in the formula (6.80) for the natural line shape function. This formula is unsymmetrical between the initial and final atomic states, since a level shift occurs for the excited state r but not for the ground state s. We should, however, calculate probabilities for transitions between atomic states with energies $E_r + \delta\tilde{E}_r$ and $E_s + \delta\tilde{E}_s$; the ground state energy of the atom when all real photons have left it is still shifted by electromagnetic interactions. Let δH_{atom} be the extension of an operator on the particle space P that is diagonal in a representation in which H_{atom} is diagonal, the diagonal elements being the Lamb-shift energies. Then

$$H = H_{\text{rad}} + (H_{\text{atom}} + \delta H_{\text{atom}}) + (H_{\text{int}} - \delta H_{\text{atom}}) \tag{6.90}$$

defines a new partitioning of the Hamiltonian which, if used in the damping formalism instead of the old partitioning (6.81), will restore the symmetry between the initial and final levels, giving

$$P(E) = \frac{\hbar A}{2\pi} \frac{1}{(E - E_{rs} - \delta\tilde{E}_r + \delta\tilde{E}_s) + \frac{1}{4}\hbar^2 A^2} \tag{6.91}$$

The ground state level shift can be calculated from ordinary stationary perturbation theory; if the ground state of H_{atom} is non-degenerate, we may assume that H has a unique normalizable ground state that passes over into that of H_0 as $e \to 0$. In the approximation used in deriving Equation (6.91), it is sufficient to calculate with the unperturbed wave functions. The level shift of the excited state cannot be obtained by elementary perturbation theory, since the excited levels lie in a continuum and have infinite degrees of degeneracy. The excited state level shifts, including the prescription for the principal value, emerge naturally from the damping formalism. A more accurate treatment of line shape problems would take into account the way in which the excited state is formed (see Heitler 1954, Power and Zienau 1959, Kroll 1964, Davidovich and Nussenzveig 1980).

REFERENCES

Arnous, E. and Heitler, W. (1953). "Theory of line-breadth phenomena". *Proc. Roy. Soc. Lond.* A **220,** 290.

Bethe, H. (1947). "The electromagnetic shift of energy levels". *Phys. Rev.* **72,** 339.

Davidovich, L. and Nussenzveig, H. M. (1980). *In* "Foundations of Radiation Theory and Quantum Electrodynamics" (ed. A. O. Barut). Plenum Press, New York and London.

Dirac, P. A. M. (1927a). "The quantum theory of the emission and absorption of radiation". *Proc. Roy. Soc. Lond. A* **114,** 243.
Dirac, P. A. M. (1927b). "The quantum theory of dispersion". *Proc. Roy. Soc. Lond. A* **114,** 710.
Dirac, P. A. M. (1958). "The Principles of Quantum Mechanics". Clarendon Press, Oxford.
Fermi, E. (1950). "Nuclear Physics". University of Chicago Press.
Healy, W. P. (1977). "A generalization of the Kramers–Heisenberg dispersion formula". *Phys. Rev. A* **16,** 1568.
Heitler, W. (1954). "The Quantum Theory of Radiation". Clarendon Press, Oxford.
Heitler, W. and Ma, S. T. (1949). "Quantum theory of radiation damping for discrete states". *Proc. Roy. Irish Acad. A* **52,** 109.
Kramers, H. A. and Heisenberg, W. (1925). "Über die Streuung von Strahlung durch atome". *Zeitschrift fur Physik* **31,** 681.
Kroll, N. (1964). *In* "Quantum Optics and Electronics" (ed. C. De Witt, A. Blandin and C. Cohen-Tannoudji). Gordon & Breach, New York.
Lamb, W. E. and Retherford, R. C. (1947). "Fine structure of the hydrogen atom by a microwave method". *Phys. Rev.* **72,** 241.
Messiah, A. (1962). "Quantum Mechanics". North Holland, Amsterdam.
Power, E. A. (1964). "Introductory Quantum Electrodynamics". Longmans, London.
Power, E. A. and Zienau, S. (1959). "Coulomb gauge in non-relativistic quantum electrodynamics and the shape of spectral lines". *Phil. Trans. Roy. Soc. Lond. A* **251,** 427.
Schweber, S. S. (1961). "An Introduction to Relativistic Quantum Field Theory". Row, Peterson, Evanston.

Chapter 7

Path-dependent Electrodynamics

7.1 Introduction

The Lagrangian corresponding to a given physical system is not uniquely determined. One method of changing the Lagrangian is to add to it the total time derivative of a function of the generalized coordinates. Let us consider, for simplicity, a classical scleronomic system with n degrees of freedom and described by a Lagrangian function $L(q, \dot{q})$. It is easy to verify that if the Lagrangian is transformed according to $\bar{L}(q, \dot{q}) = L(q, \dot{q}) - \mathrm{d}W/\mathrm{d}t$ for some function $W(q)$, then the Euler–Lagrange equations are unaltered. The addition of a total time derivative to the Lagrangian induces a canonical transformation in the phase space of the system. A non-singular transformation from canonical variables (p, q) to new variables $(\bar{p}, \bar{q})$ is said to be canonical if Hamilton's equations are covariant under it, the new Hamiltonian function being defined by $\bar{H}(\bar{p}, \bar{q}) = H(p, q)$. Here and in the following we adopt the alias or passive interpretation of the transformation, so that (p, q) and $(\bar{p}, \bar{q})$ refer to the same point in phase space but in two different coordinate systems. The necessary and sufficient condition for a coordinate transformation in phase space to be canonical is (Lanczos 1970) that the differential form $p_i\,\delta q_i$ should change by at most a perfect differential, i.e. that $\bar{p}_i\,\delta\bar{q}_i = p_i\,\delta q_i - \delta W$ for some function $W(p, q)$. The addition of the term $-\delta W$ amounts only to the addition of a vanishing boundary term to the variation of the canonical integral, since the value of the Hamiltonian is invariant and all variations vanish at the end points. Thus Hamilton's principle holds in the new coordinate system if and only if it holds in the old, and the condition for this is that the canonical equations be covariant. For those canonical transformations induced by adding a total time derivative to the Lagrangian, we have $\bar{p}_i = \partial\bar{L}/\partial q_i = p_i - \partial W/\partial q_i$ and, since $\bar{q}_i = q_i$, $\bar{p}_i\,\delta\bar{q}_i = p_i\,\delta q_i - \delta W$. In this case W depends on q but not on p. The equality, at any point in phase space, of the Hamiltonian $\bar{H}$ derived from the Lagrangian $\bar{L}$ and the Hamiltonian H derived from the Lagrangian L may readily be verified. It should be noted, however, that, unless the transformation corresponds to a symmetry of the system, the functional form of the Hamiltonian changes under a canonical transformation.

In this chapter we consider a particular canonical transformation in the phase space for the complete system of electromagnetic field and charged particles. It will be shown in Section 7.3 that carrying out a gauge transformation from the potentials of the Coulomb gauge to those of the line integral gauge introduced in Section 2.4 results in the addition of a total time derivative to the Lagrangian (Healy 1979, 1980). The associated canonical transformation (Power and Zienau 1959, Woolley 1975a, Power 1978), to be performed in Section 7.4, leads to the so-called "multipolar" form of the Hamiltonian in which the interaction term may be expanded as a series of electric and magnetic multipoles (Göppert–Mayer 1931, Richards 1948, Fiutak 1963, de Groot 1969, Atkins and Woolley 1970). In the multipolar form of the Hamiltonian, the field conjugate to the vector potential in the Coulomb gauge is proportional to the transverse displacement vector and not, as in the "minimal-coupling" form used in Chapters 3 to 6, to the transverse electric field, and similarly the canonical momenta of the charged particles differ in the two cases. The division of the Hamiltonian in its multipolar form into unperturbed and interaction parts will be seen to be intrinsically path-dependent, the path dependence of the potentials being merely transferred to the polarization and magnetization fields for the aggregate of charged particles. We begin, in Section 7.2, with a discussion of the polarization and magnetization fields. Some applications of the multipolar formalism are given in Section 7.5.

PROBLEMS

(1) For a scleronomic system with n degrees of freedom, a Lagrangian point transformation is a coordinate change $q \rightarrow \bar{q}$ in configuration space such that the Jacobian

$$\partial(\bar{q}_1, \ldots, \bar{q}_n)/\partial(q_1, \ldots, q_n)$$

is non-zero, the Lagrangian function transforming according to $\bar{L}(\bar{q}, \dot{\bar{q}}) = L(q, \dot{q})$.

(i) Show by direct substitution that the Euler–Lagrange equations are covariant under Lagrangian point transformations.

(ii) Show that the set of all Lagrangian point transformations for a given configuration space forms a non-Abelian group.

(iii) Show that a Lagrangian point transformation in configuration space induces a canonical transformation in phase space.

(2) Show that, provided the determinants of the Hessian matrices associated with the generating functions $W_1(q, \bar{q})$, $W_2(q, \bar{p})$, $W_3(p, \bar{q})$ and $W_4(p, \bar{p})$ are non-zero, each of the following four sets of equations implicitly defines a canonical transformation:

(i) $p_i = \partial W_1/\partial q_i \qquad \bar{p}_i = -\partial W_1/\partial \bar{q}_i$

(ii) $p_i = \partial W_2/\partial q_i \qquad \bar{q}_i = \partial W_2/\partial \bar{p}_i$

(iii) $q_i = -\partial W_3/\partial p_i \qquad \bar{p}_i = -\partial W_3/\partial \bar{q}_i$

(iv) $q_i = -\partial W_4/\partial p_i \qquad \bar{q}_i = \partial W_4/\partial \bar{p}_i$

(3) (i) Show that the identity transformation can be obtained from either of the generating functions

$$W_2(q, \bar{p}) = q_i\bar{p}_i \qquad W_3(p, \bar{q}) = -p_i\bar{q}_i$$

(ii) Show that the transformation $\bar{p}_1 = q_1$, $\bar{q}_1 = -p_1$, $\bar{p}_2 = p_2$, $\bar{q}_2 = q_2$ is *not* derivable from a generating function of the type W_1, W_2, W_3 or W_4, but is nevertheless a canonical transformation.

(4) Prove that if $W, \Omega_1, \ldots, \Omega_k$ are $k+1$ functions of the $2n$ variables $(q, \bar{q})$, where $k \leqslant n$, and if $(\bar{p}, \bar{q}, \lambda_1, \ldots, \lambda_k)$ are defined in terms of (p, q) by the $2n+k$ equations

$$\Omega_i(q, \bar{q}) = 0 \qquad (i = 1, \ldots, k)$$

$$p_i = \frac{\partial W}{\partial q_i} + \lambda_1 \frac{\partial \Omega_1}{\partial q_i} + \ldots + \lambda_k \frac{\partial \Omega_k}{\partial q_i} \qquad (i = 1, \ldots, n)$$

$$\bar{p}_i = -\frac{\partial W}{\partial \bar{q}_i} - \lambda_1 \frac{\partial \Omega_1}{\partial \bar{q}_i} - \ldots - \lambda_k \frac{\partial \Omega_k}{\partial \bar{q}_i} \qquad (i = 1, \ldots, n)$$

then the transformation from (p, q) to $(\bar{p}, \bar{q})$ is a canonical transformation.

(The converse is also true—an arbitrary canonical transformation can be expressed in this form (Whittaker 1961). The equations $\Omega_i = 0$ give those independent relations (if any) that exist among the coordinates only (without involving the momenta), and the λ_i are Lagrangian multipliers.)

7.2 Polarization and magnetization fields

The problem of representing microscopic charge and current densities in terms of polarization and magnetization fields has been of importance in electrodynamics since Lorentz first presented his electron theory. The use of polarization and magnetization fields permits the division of the total charge density associated with an aggregate of point particles into "true" (ionic or free electronic) and polarization charge densities, and of the total current density into true, polarization and magnetization current densities. Such a division is useful in formulating the atomic field equations which lead, through a suitable statistical averaging procedure, to the macroscopic Maxwell equations for bulk matter (de Groot 1969). The true or atomic charge and current densities ρ_{true} and $\boldsymbol{j}_{\text{true}}$ are (paradoxically) those of a hypothetical point particle coincident with an arbitrarily chosen reference point and having charge equal to the total charge of the aggregate. The atomic field equations that involve sources relate the true charge and current densities to the electric displacement vector $\boldsymbol{d}$ and magnetic field $\boldsymbol{h}$:

$$\nabla \cdot \boldsymbol{d} = 4\pi\rho_{\text{true}} \tag{7.1}$$

$$\nabla \times \boldsymbol{h} = \frac{4\pi}{c}\boldsymbol{j}_{\text{true}} + \frac{1}{c}\dot{\boldsymbol{d}} \tag{7.2}$$

where $\boldsymbol{d} = \boldsymbol{e} + 4\pi\boldsymbol{p}$ and $\boldsymbol{h} = \boldsymbol{b} - 4\pi\boldsymbol{m}$. It follows that the microscopic polarization and magnetization fields $\boldsymbol{p}$ and $\boldsymbol{m}$ must satisfy the equations

$$\begin{aligned}\rho &= Q\,\delta(\boldsymbol{x} - \boldsymbol{R}) - \nabla \cdot \boldsymbol{p} \\ &\equiv \rho_{\text{true}} + \rho_{\text{polarization}}\end{aligned} \tag{7.3}$$

$$\begin{aligned}\boldsymbol{j} &= Q\dot{\boldsymbol{R}}\,\delta(\boldsymbol{x} - \boldsymbol{R}) + \dot{\boldsymbol{p}} + c\nabla \times \boldsymbol{m} \\ &\equiv \boldsymbol{j}_{\text{true}} + \boldsymbol{j}_{\text{polarization}} + \boldsymbol{j}_{\text{magnetization}}\end{aligned} \tag{7.4}$$

where $\boldsymbol{R}$ is the reference point and Q the total charge of the aggregate. In this section we give the general solution to the problem of finding polarization and magnetization fields to represent a given classical microscopic charge-current distribution (see Healy 1977b, 1978, where also references to previous work may be found).

We consider an aggregate of charged particles moving in a given inertial frame of reference. A particle labelled α is supposed, as before, to have charge e_α and position vector $\boldsymbol{q}_\alpha$ and may be either an electron or a nucleus, while the aggregate itself may be a netural atom or molecule, an ion, or even a single electron. (The fixed-nuclei approximation is now not necessarily being used.) These particles give rise at the field point $\boldsymbol{x}$ and the time t to the microscopic charge and current densities ρ and $\boldsymbol{j}$ defined in Equations (2.5) and (2.6). To analyse the total charge and current densities into true and bound contributions, it is necessary to introduce the privileged central or reference point $\boldsymbol{R}$ with the motion of which that of the aggregate as a whole is to be identified. This point could be, for instance, the centre of mass or, for a system that is not overall electrically neutral, the centre of charge, but its precise specification will not be needed yet. Indeed, from the point of view of the purely kinematical theory, which alone concerns us in this section, the reference point $\boldsymbol{R}$ could be taken to trace out an arbitrary smooth trajectory unconnected with any motion of the particles in the aggregate. We next choose, for each instant of time t and each particle α, a spatial curve C_α starting at $\boldsymbol{R}$ and ending at $\boldsymbol{q}_\alpha$. The length and shape of these curves could alter in an infinity of different ways as the end points are carried along by the $\boldsymbol{q}_\alpha$ and $\boldsymbol{R}$ in their motion; we assume this to happen, however, in a definite and continuous manner. The exact meaning to be attached to the instantaneous velocity of an intermediate point $\boldsymbol{x}'$ of any of the C_α is explained in Appendix B. Polarization and magnetization fields associated with a given reference point $\boldsymbol{R}$ and given curves C_α may be defined by

$$\boldsymbol{p}(\boldsymbol{x}, t) = \sum_\alpha e_\alpha \int_{C_\alpha} \mathrm{d}\boldsymbol{x}'\, \delta(\boldsymbol{x}-\boldsymbol{x}') \tag{7.5}$$

$$\boldsymbol{m}(\boldsymbol{x}, t) = \sum_\alpha \frac{e_\alpha}{c} \int_{C_\alpha} \mathrm{d}\boldsymbol{x}' \times \dot{\boldsymbol{x}}'\, \delta(\boldsymbol{x}-\boldsymbol{x}') \tag{7.6}$$

We have first that

$$\begin{aligned}\nabla \cdot \boldsymbol{p} &= -\sum_\alpha e_\alpha \int_{C_\alpha} \mathrm{d}\boldsymbol{x}' \cdot \nabla' \delta(\boldsymbol{x}-\boldsymbol{x}') \\ &= -\sum_\alpha e_\alpha\, \delta(\boldsymbol{x}-\boldsymbol{q}_\alpha) + Q\, \delta(\boldsymbol{x}-\boldsymbol{R})\end{aligned} \tag{7.7}$$

which is equivalent to Equation (7.3). We have also from Theorem 7 of Appendix B that

$$\begin{aligned}\nabla \times \boldsymbol{m} &= \nabla \times \sum_\alpha \frac{e_\alpha}{c} \int_{C_\alpha} \mathrm{d}\boldsymbol{x}' \times \dot{\boldsymbol{x}}'\, \delta(\boldsymbol{x}-\boldsymbol{x}') \\ &= \sum_\alpha \frac{e_\alpha}{c} \dot{\boldsymbol{q}}_\alpha\, \delta(\boldsymbol{x}-\boldsymbol{q}_\alpha) - \frac{Q}{c} \dot{\boldsymbol{R}}\, \delta(\boldsymbol{x}-\boldsymbol{R}) \\ &\quad - \frac{1}{c} \frac{\partial}{\partial t} \sum_\alpha e_\alpha \int_{C_\alpha} \mathrm{d}\boldsymbol{x}'\, \delta(\boldsymbol{x}-\boldsymbol{x}')\end{aligned} \tag{7.8}$$

which, in view of the definition (7.5), is equivalent to Equation (7.4).

The expressions (7.5) and (7.6) for the polarization and magnetization fields can be interpreted as follows. If to any real charge e_α we add two fictitious charges e_α and $-e_\alpha$, both located at a point $\boldsymbol{x}'$ in the immediate neighbourhood of $\boldsymbol{q}_\alpha$ and both moving with a velocity $\dot{\boldsymbol{x}}'$ that is infinitesimally close to $\dot{\boldsymbol{q}}_\alpha$, then this does not alter the system physically, since the total added charges and currents are zero and have no mass associated with them. The real moving charge e_α and the fictitious moving charge

$-e_\alpha$ form an electric dipole of moment $e_\alpha\,\mathrm{d}\boldsymbol{x}'$ and a magnetic dipole of moment $(e_\alpha/c)\,\mathrm{d}\boldsymbol{x}'\times\dot{\boldsymbol{x}}'$, where $\mathrm{d}\boldsymbol{x}'=\boldsymbol{q}_\alpha-\boldsymbol{x}'$. The remaining fictitious charge e_α located at and moving with the point $\boldsymbol{x}'$ can be regarded as the original charge displaced and given a velocity increment, the addition of the electric and magnetic dipoles compensating for the change. This charge can now be given a further infinitesimal displacement and a further infinitesimal velocity increment, provided that further compensating dipoles are added to the system. Continuing in this way we may shift the actual charge-current density at $\boldsymbol{q}_\alpha$ in a continuously varying manner along a moving curve C_α (taken in its negative sense) until the reference point $\boldsymbol{R}$ is reached. If this is done for all particles α, the total charge-current density becomes concentrated at the reference point and gives rise to the true charge-current density there. The volume densities of the compensating electric and magnetic point dipoles distributed along the curves C_α are then just the polarization and magnetization fields (7.5) and (7.6). The charge-current densities associated with these distributions together with the true charge-current density make up the total, real, charge-current density, since the system has not been changed physically; this is the content of Equations (7.3) and (7.4). It should be emphasized that moving the charges and currents along the curves C_α and adding the compensating dipoles are, in the language of the Calculus of Variations, virtual processes, i.e. they are imagined to be carried out instantaneously, and for each time t. The virtual work done by external fields $\boldsymbol{E}$ and $\boldsymbol{B}$ on the charges and currents on being moved, in the negative sense, along the curves C_α at a given time t is exactly balanced by the work that must be done against $\boldsymbol{E}$ and $\boldsymbol{B}$ to create the dipole distribution at the same time t.

We have seen that it is possible to find particular polarization and magnetization fields that enable us to represent the total charge and current densities as sums of true and bound contributions. These fields are not unique, however, as is evident first from the freedom in the choice of $\boldsymbol{R}$ and the curves C_α. We now determine the general class of polarization and magnetization fields that reproduce, when associated with a given reference point, the total charge and current densities through Equations (7.3) and (7.4). Let us suppose we have a reference point $\boldsymbol{R}_1$ and fields $\boldsymbol{p}^{(1)}$ and $\boldsymbol{m}^{(1)}$ that satisfy these equations. (It is not now assumed that $\boldsymbol{p}^{(1)}$ and $\boldsymbol{m}^{(1)}$ are defined by line integrals.) Let us suppose also that with the reference point $\boldsymbol{R}_2$, which need not coincide with $\boldsymbol{R}_1$, are associated the fields $\boldsymbol{p}^{(2)}$ and $\boldsymbol{m}^{(2)}$, and that these too satisfy Equations (7.3) and (7.4). To exhibit the connection between $\boldsymbol{p}^{(1)}$ and $\boldsymbol{m}^{(1)}$ and $\boldsymbol{p}^{(2)}$ and $\boldsymbol{m}^{(2)}$, we imagine the reference points, if these are distinct, to be joined by a curve C_{12} starting at $\boldsymbol{R}_1$ and ending at $\boldsymbol{R}_2$ and which moves along with these points in the manner described in Appendix B. If $\boldsymbol{R}_1$ and

$\boldsymbol{R}_2$ always coincide, the curve degenerates into a point and any line integral taken over it is zero. Now if Equation (7.3) is satisfied simultaneously by $\boldsymbol{p}^{(1)}$ and $\boldsymbol{R}_1$ and by $\boldsymbol{p}^{(2)}$ and $\boldsymbol{R}_2$, then

$$\begin{aligned}\nabla\cdot(\boldsymbol{p}^{(2)}-\boldsymbol{p}^{(1)}) &= Q\{\delta(\boldsymbol{x}-\boldsymbol{R}_2)-\delta(\boldsymbol{x}-\boldsymbol{R}_1)\}\\ &= -Q\nabla\cdot\int_{C_{12}} \mathrm{d}\boldsymbol{x}'\,\delta(\boldsymbol{x}-\boldsymbol{x}')\end{aligned} \tag{7.9}$$

and if Equation (7.4) is satisfied simultaneously by $\boldsymbol{p}^{(1)}$, $\boldsymbol{m}^{(1)}$ and $\boldsymbol{R}_1$ and by $\boldsymbol{p}^{(2)}$, $\boldsymbol{m}^{(2)}$ and $\boldsymbol{R}_2$, then

$$\begin{aligned}c\nabla\times(\boldsymbol{m}^{(2)}-\boldsymbol{m}^{(1)})+\dot{\boldsymbol{p}}^{(2)}-\dot{\boldsymbol{p}}^{(1)} &= -Q\{\dot{\boldsymbol{R}}_2\,\delta(\boldsymbol{x}-\boldsymbol{R}_2)-\dot{\boldsymbol{R}}_1\,\delta(\boldsymbol{x}-\boldsymbol{R}_1)\}\\ &= -Q\left\{\frac{\partial}{\partial t}\int_{C_{12}} \mathrm{d}\boldsymbol{x}'\,\delta(\boldsymbol{x}-\boldsymbol{x}')+\nabla\times\int_{C_{12}} \mathrm{d}\boldsymbol{x}'\times\dot{\boldsymbol{x}}'\delta(\boldsymbol{x}-\boldsymbol{x}')\right\}\end{aligned} \tag{7.10}$$

as follows from Theorem 7 of Appendix B. For Equations (7.9) and (7.10) to be valid, we must have

$$\boldsymbol{p}^{(2)} = \boldsymbol{p}^{(1)} - Q\int_{C_{12}} \mathrm{d}\boldsymbol{x}'\,\delta(\boldsymbol{x}-\boldsymbol{x}')+\nabla\times\boldsymbol{v}^{(12)} \tag{7.11}$$

$$\boldsymbol{m}^{(2)} = \boldsymbol{m}^{(1)} - \frac{Q}{c}\int_{C_{12}} \mathrm{d}\boldsymbol{x}'\times\dot{\boldsymbol{x}}'\,\delta(\boldsymbol{x}-\boldsymbol{x}')-\frac{1}{c}\dot{\boldsymbol{v}}^{(12)}-\nabla s^{(12)} \tag{7.12}$$

for some vector field $\boldsymbol{v}^{(12)}(\boldsymbol{x}, t)$ and some scalar field $s^{(12)}(\boldsymbol{x}, t)$. These equations give the relations that must exist between two sets of polarization and magnetization fields and their associated reference points. Conversely, if fields $\boldsymbol{p}^{(1)}$ and $\boldsymbol{m}^{(1)}$ associated with the reference point $\boldsymbol{R}_1$ satisfy Equations (7.3) and (7.4), then for arbitrary differentiable fields $\boldsymbol{v}^{(12)}$ and $s^{(12)}$ and an arbitrary smoothly moving curve C_{12} the fields $\boldsymbol{p}^{(2)}$ and $\boldsymbol{m}^{(2)}$ defined by Equations (7.11) and (7.12) and associated with the reference point $\boldsymbol{R}_2$ will also satisfy Equations (7.3) and (7.4). It may readily be shown that the relation between $\boldsymbol{p}^{(1)}$, $\boldsymbol{m}^{(1)}$ and $\boldsymbol{R}_1$ on the one hand, and $\boldsymbol{p}^{(2)}$, $\boldsymbol{m}^{(2)}$ and $\boldsymbol{R}_2$ on the other, that is implied by Equations (7.11) and (7.12), is an equivalence relation in the mathematical sense, i.e. it is reflexive, symmetric and transitive.

The transformation (7.11) and (7.12) may be written in terms of a single vector field $\boldsymbol{u}^{(12)}$ defined through its transverse and longitudinal parts by the equations

$$\boldsymbol{u}^{(12)\perp}(\boldsymbol{x}, t) = \boldsymbol{v}^{(12)\perp}(\boldsymbol{x}, t) \tag{7.13}$$

and

$$\boldsymbol{u}^{(12)\parallel}(\boldsymbol{x}, t) = \boldsymbol{v}^{(12)\parallel}(\boldsymbol{x}, t) + c\int^{t} \nabla s^{(12)}(\boldsymbol{x}, t')\,\mathrm{d}t' \tag{7.14}$$

respectively. For then

$$\nabla \times \boldsymbol{v}^{(12)} = \nabla \times \boldsymbol{u}^{(12)} \tag{7.15}$$

and

$$\frac{1}{c}\dot{\boldsymbol{v}}^{(12)} + \nabla s^{(12)} = \frac{1}{c}\dot{\boldsymbol{u}}^{(12)} \tag{7.16}$$

In particular, if the reference points $\boldsymbol{R}_1$ and $\boldsymbol{R}_2$ coincide and are fixed in space, then the transformation reduces to

$$\boldsymbol{p}^{(2)} = \boldsymbol{p}^{(1)} + \nabla \times \boldsymbol{u}^{(12)} \tag{7.17}$$

$$\boldsymbol{m}^{(2)} = \boldsymbol{m}^{(1)} - \frac{1}{c}\dot{\boldsymbol{u}}^{(12)} \tag{7.18}$$

This is the form of the transformation that has been given by Woolley (1975a, b).

The transformation equations (7.11) and (7.12) can be verified and the vector and scalar fields $\boldsymbol{v}^{(12)}$ and $s^{(12)}$ can be obtained explicitly for the particular case of line integral polarization and magnetization fields. Let $\boldsymbol{p}^{(1)}$ and $\boldsymbol{m}^{(1)}$ be defined in terms of a reference point $\boldsymbol{R}_1$ and integration paths $C_\alpha^{(1)}$ and let $\boldsymbol{p}^{(2)}$ and $\boldsymbol{m}^{(2)}$ be defined in terms of a reference point $\boldsymbol{R}_2$ and integration paths $C_\alpha^{(2)}$. If $\boldsymbol{R}_1$ and $\boldsymbol{R}_2$ are distinct, we again suppose them joined by a curve C_{12} and choose for each particle α a surface Σ_α bounded by the curves C_{12}, $C_\alpha^{(1)}$ and $C_\alpha^{(2)}$ and which moves along in a definite and continuous manner with them. The total boundary curve Γ_α of Σ_α will be taken to be traced out in its positive sense by going from $\boldsymbol{R}_1$ to $\boldsymbol{R}_2$ along C_{12}, then from $\boldsymbol{R}_2$ to $\boldsymbol{q}_\alpha$ along $C_\alpha^{(2)}$ and finally from $\boldsymbol{q}_\alpha$ back to $\boldsymbol{R}_1$ in the negative sense along $C_\alpha^{(1)}$. If $\boldsymbol{R}_1$ and $\boldsymbol{R}_2$ always coincide, Γ_α consists of $C_\alpha^{(2)}$ and $C_\alpha^{(1)}$ only and any line integral over C_{12} is zero. The sense of Σ_α is determined from that of Γ_α by the right-hand rule in a right-handed frame and the left-hand rule in a left-handed frame. We then have from Equation (7.5) and Theorem 3 of Appendix B

$$\begin{aligned}\boldsymbol{p}^{(2)} - \boldsymbol{p}^{(1)} + Q\int_{C_{12}} \mathrm{d}\boldsymbol{x}'\,\delta(\boldsymbol{x}-\boldsymbol{x}') &= \sum_\alpha e_\alpha \oint_{\Gamma_\alpha} \mathrm{d}\boldsymbol{x}'\,\delta(\boldsymbol{x}-\boldsymbol{x}') \\ &= \sum_\alpha e_\alpha \iint_{\Sigma_\alpha} \mathrm{d}\boldsymbol{S}' \times \nabla'\,\delta(\boldsymbol{x}-\boldsymbol{x}') \\ &= \nabla \times \sum_\alpha e_\alpha \iint_{\Sigma_\alpha} \mathrm{d}\boldsymbol{S}'\,\delta(\boldsymbol{x}-\boldsymbol{x}')\end{aligned} \tag{7.19}$$

and from Equation (7.6) and Theorems 4 and 8 of Appendix B

$$\boldsymbol{m}^{(2)}-\boldsymbol{m}^{(1)}+\frac{Q}{c}\int_{C_{12}} \mathrm{d}\boldsymbol{x}'\times\dot{\boldsymbol{x}}'\,\delta(\boldsymbol{x}-\boldsymbol{x}')$$

$$=\sum_{\alpha}\frac{e_\alpha}{c}\oint_{\Gamma_\alpha} \mathrm{d}\boldsymbol{x}'\times\dot{\boldsymbol{x}}'\,\delta(\boldsymbol{x}-\boldsymbol{x}')$$

$$=\sum_{\alpha}\frac{e_\alpha}{c}\iint_{\Sigma_\alpha}(\mathrm{d}\boldsymbol{S}'\times\nabla')\times\{\dot{\boldsymbol{x}}'\,\delta(\boldsymbol{x}-\boldsymbol{x}')\}$$

$$=-\frac{1}{c}\frac{\partial}{\partial t}\sum_{\alpha}e_\alpha\iint_{\Sigma_\alpha}\mathrm{d}\boldsymbol{S}'\,\delta(\boldsymbol{x}-\boldsymbol{x}')-\nabla\sum_{\alpha}\frac{e_\alpha}{c}\iint_{\Sigma_\alpha}\mathrm{d}\boldsymbol{S}'\,.\,\dot{\boldsymbol{x}}'\,\delta(\boldsymbol{x}-\boldsymbol{x}') \quad (7.20)$$

These equations are evidently of the same form as Equations (7.11) and (7.12) with

$$\boldsymbol{v}^{(12)}(\boldsymbol{x},t)=\sum_{\alpha}e_\alpha\iint_{\Sigma_\alpha}\mathrm{d}\boldsymbol{S}'\,\delta(\boldsymbol{x}-\boldsymbol{x}') \quad (7.21)$$

$$s^{(12)}(\boldsymbol{x},t)=\sum_{\alpha}\frac{e_\alpha}{c}\iint_{\Sigma_\alpha}\mathrm{d}\boldsymbol{S}'\,.\,\dot{\boldsymbol{x}}'\,\delta(\boldsymbol{x}-\boldsymbol{x}') \quad (7.22)$$

so that the general transformation rule is obeyed.

PROBLEMS

(1) Suppose that for given polarization and magnetization fields $\boldsymbol{p}^{(1)}$ and $\boldsymbol{m}^{(1)}$ and $\boldsymbol{p}^{(2)}$ and $\boldsymbol{m}^{(2)}$ associated with given reference points $\boldsymbol{R}_1$ and $\boldsymbol{R}_2$, respectively, Equations (7.11) and (7.12) are satisfied by vector and scalar fields $\boldsymbol{v}^a$ and s^a and $\boldsymbol{v}^b$ and s^b associated with curves C_{12}^a and C_{12}^b, respectively. Show that $\boldsymbol{v}^b$ and s^b must be related to $\boldsymbol{v}^a$ and s^a through the transformation

$$\boldsymbol{v}^b=\boldsymbol{v}^a-Q\iint_{S^{ab}}\mathrm{d}\boldsymbol{S}'\,\delta(\boldsymbol{x}-\boldsymbol{x}')-\nabla f$$

$$s^a=s^a-\frac{Q}{c}\iint_{S^{ab}}\mathrm{d}\boldsymbol{S}'\,.\,\dot{\boldsymbol{x}}'\,\delta(\boldsymbol{x}-\boldsymbol{x}')+\frac{1}{c}\frac{\partial f}{\partial t}$$

where S^{ab} is a surface bounded by and moving with the curves C_{12}^a and C_{12}^b (the latter being taken in its negative sense) and f is a function of $\boldsymbol{x}$ and t. Hence show that the relation between any ($\boldsymbol{v}^a$, s^a, C_{12}^a) and ($\boldsymbol{v}^b$, s^b, C_{12}^b) that satisfy Equations (7.11) and (7.12) for given ($\boldsymbol{p}^{(1)}$, $\boldsymbol{m}^{(1)}$, $\boldsymbol{R}_1$) and ($\boldsymbol{p}^{(2)}$, $\boldsymbol{m}^{(2)}$, $\boldsymbol{R}_2$) is an equivalence relation.

(2) Use Theorems 5 and 9 of Appendix B to (i) investigate the dependence of the function f of the previous question on the choice, for given boundary curves C_{12}^a and C_{12}^b, of the surface S^{ab} and (ii) show that f can be expressed explicitly as a volume integral in the case where both $\boldsymbol{v}^a$ and s^a and $\boldsymbol{v}^b$ and s^b are defined by surface integrals, as in Equations (7.21) and (7.22).

7.3 Line integral Lagrangians

The Lagrangian L for the coupled systems of field and particles that was given in Section 3.4 is a function of the particle coordinates and velocities $\boldsymbol{q}_\alpha$ and $\dot{\boldsymbol{q}}_\alpha$ and a functional of the transverse vector potential $\boldsymbol{a}$ and the corresponding velocity $\dot{\boldsymbol{a}}$. In the application of Hamilton's principle to this Lagrangian, the Coulomb gauge condition was imposed on the vector potential $\boldsymbol{a}$ and the scalar potential ϕ was treated as a prescribed function of the particle coordinates. The Euler–Lagrange equations for the particle variables were shown to be equivalent to Newton's law with the Lorentz force and those for the field variables to the Maxwell–Lorentz equations involving sources, the source-free equations being implicit in the relation of the electric and magnetic induction fields to the potentials. The effect of gauge transformations on the Lagrangian may be exhibited by first adding and subtracting the Coulomb energy (with infinite self-energy terms being assumed to be discarded) to obtain

$$L = L_0 + L_1 \tag{7.23}$$

where

$$L_0 = \frac{1}{8\pi}\iiint (\boldsymbol{e}^2 - \boldsymbol{b}^2)\,\mathrm{d}^3x + \frac{1}{2}\sum_\alpha m_\alpha \dot{\boldsymbol{q}}_\alpha^2 \tag{7.24}$$

and

$$L_1 = \iiint \left(\frac{1}{c}\,\boldsymbol{j}\,.\,\boldsymbol{a} - \rho\phi\right)\mathrm{d}^3x \tag{7.25}$$

It should be noted that L_0 is not the sum of L_{par} and L_{rad} given by Equations (3.60) and (3.61) and that L_1 differs from L_{int} given by Equation (3.62). Under the substitution

$$\phi \to \bar{\phi} = \phi + \frac{1}{c}\dot{\chi} \qquad \boldsymbol{a} \to \bar{\boldsymbol{a}} = \boldsymbol{a} - \nabla\chi \tag{7.26}$$

which corresponds to a gauge transformation with gauge function χ, L_0 is invariant ($\bar{L}_0 = L_0$) and

$$L_1 \to \bar{L}_1 \equiv \iiint \left(\frac{1}{c}\boldsymbol{j}\,.\,\bar{\boldsymbol{a}} - \rho\bar{\phi}\right) \mathrm{d}^3x$$

$$= L_1 - \frac{1}{c}\iiint (\boldsymbol{j}\,.\,\nabla\chi + \rho\dot{\chi})\,\mathrm{d}^3x \qquad (7.27)$$

We shall continue to regard the $\boldsymbol{q}_\alpha$ and $\boldsymbol{a}$ as the Lagrangian coordinates for the system and shall suppose that χ is a prescribed function or functional of these. Then if $\chi(\boldsymbol{x}, t)$ depends on the $\boldsymbol{q}_\alpha$ and $\boldsymbol{a}$ only through their values at time t, the Lagrangians $\bar{L}$ and L are equivalent, i.e. they yield equivalent equations of motion for the system. For the continuity equation (2.7) implies that

$$\frac{\mathrm{d}}{\mathrm{d}t}\iiint \rho\chi\,\mathrm{d}^3x = \iiint (-\chi\nabla\,.\,\boldsymbol{j} + \rho\dot{\chi})\,\mathrm{d}^3x$$

$$= \iiint (\boldsymbol{j}\,.\,\nabla\chi + \rho\dot{\chi})\,\mathrm{d}^3x \qquad (7.28)$$

where the last step follows from an integration by parts. It is legitimate to use the continuity equation to reduce the Lagrangian, since this equation is a consequence of the functional dependence of the charge and current densities on the particle coordinates and velocities (Equations (2.5) and (2.6)) and thus holds as well for the varied as for the natural motion of the system. From Equations (7.27) and (7.28) we obtain

$$\bar{L} = L - \frac{1}{c}\frac{\mathrm{d}}{\mathrm{d}t}\iiint \rho\chi\,\mathrm{d}^3x \qquad (7.29)$$

so that $\bar{L}$ and L differ by the total time derivative of a quantity that does not depend on the Lagrangian velocities. The variations of the action integral (subject to the Coulomb gauge condition) must then be the same,

$$\delta\int_{t_1}^{t_2} \bar{L}\,\mathrm{d}t = \delta\int_{t_1}^{t_2} L\,\mathrm{d}t \qquad (7.30)$$

since the variations of the coordinates vanish at t_1 and t_2. (The variations of the velocities do not necessarily vanish there.) It follows that the Euler–Lagrange equations derived from $\bar{L}$ are equivalent to those derived from L. This is because the Euler–Lagrange equations are both the necessary and sufficient conditions for Hamilton's principle to hold.

We consider now the Lagrangian $\bar{L}$ for the case in which $\bar{\boldsymbol{a}}$ and $\bar{\phi}$ are the line integral vector and scalar potentials discussed in Section 2.4, the original potentials $\boldsymbol{a}$ and ϕ being those of the Coulomb gauge. $\bar{L}$ will then be referred to as a line-integral Lagrangian. The original Lagrangian L has no explicit time dependence, since the interacting system of field and particles is scleronomic. If the same is to be true of $\bar{L}$, then the motion of the reference point $\boldsymbol{R}$ and of the curves $C_{\boldsymbol{x}}$ that define the potentials must not be externally prescribed but must be a consequence of the motion of the system. We assume that there is a functional relation between the curves and the dynamical variables and that this is time-independent and involves the coordinates only, so that derivatives higher than the first are not introduced into the Lagrangian. The reference point can still be chosen in many ways and need not be stationary nor at infinity nor coincident with a material particle. If, however, the reference point that is used to define the line-integral potentials is also used to define the true charge and current densities, as will be done here, then it is natural to suppose that $\boldsymbol{R}$ always remains in the vicinity of, and thus moves with, the aggregate of charged particles. For a given choice of reference point, the curves $C_{\boldsymbol{x}}$ can also be chosen in many ways. The simplest approach is to use the straight lines joining $\boldsymbol{R}$ to the field points $\boldsymbol{x}$. We could also use paths that are not necessarily rectilinear, e.g. those defined parametrically by $\boldsymbol{x}'=f(u)\boldsymbol{x}+F(u)\boldsymbol{R}$, $u_1 \leqslant u \leqslant u_2$, where $f(u)$ and $F(u)$ are continuously differentiable functions such that $f(u_1)=0=F(u_2)$ and $F(u_1)=1=f(u_2)$. In the following, we shall continue to specify the curves $C_{\boldsymbol{x}}$ by the functions $\boldsymbol{x}'(u, \boldsymbol{x}, t)$ used in Section 2.4, but shall understand that the time dependence is determined dynamically in the way just described.

If we substitute the line integral potentials of Equations (2.29) and (2.30) in the first line of Equation (7.27) and use the expressions (2.5) and (2.6) for ρ and $\boldsymbol{j}$, we obtain

$$\bar{L}_1 = \frac{Q}{c}\dot{\boldsymbol{R}} \cdot \boldsymbol{a}(\boldsymbol{R}, t) - Q\phi(\boldsymbol{R}, t) + \sum_\alpha e_\alpha \int_{C_{\boldsymbol{q}_\alpha}} \boldsymbol{e}(\boldsymbol{x}', t) \cdot \mathrm{d}\boldsymbol{x}'$$
$$+ \sum_\alpha \frac{e_\alpha}{c} \int_{C_{\boldsymbol{q}_\alpha}} \left(\dot{\boldsymbol{x}}' + \frac{\partial \boldsymbol{x}'}{\partial q_{\alpha i}} \dot{q}_{\alpha i}\right) \cdot \boldsymbol{b}(\boldsymbol{x}', t) \times \mathrm{d}\boldsymbol{x}' \quad (7.31)$$

The curve $C_{\boldsymbol{q}_\alpha}$ is, as the notation indicates, merely that curve $C_{\boldsymbol{x}}$ for which $\boldsymbol{x}=\boldsymbol{q}_\alpha$, the value of $\boldsymbol{q}_\alpha$ being determined from the trajectory of particle α and the time t. The points of $C_{\boldsymbol{q}_\alpha}$ are given by $\boldsymbol{x}'=\boldsymbol{x}'(u, \boldsymbol{q}_\alpha, t)$. The endpoint corresponding to $u=u_2$ is at the fixed point $\boldsymbol{q}_\alpha$ and the velocity $\dot{\boldsymbol{x}}'$ of any point is obtained by differentiating $\boldsymbol{x}'$ with respect to the third argument only. We now define for each α a curve C_α selected from the curves $C_{\boldsymbol{x}}$ by the motion of particle α. The points $\boldsymbol{x}'$ of C_α are to be given by $\boldsymbol{x}'=\boldsymbol{x}'[u, \boldsymbol{q}_\alpha(t), t]$ and the velocities $\dot{\boldsymbol{x}}'$ are to be given (for various

u) by the total time derivative of this function with respect to t. For fixed t, C_α coincides with $C_{\boldsymbol{q}_\alpha}$. However, since

$$\frac{\mathrm{d}}{\mathrm{d}t}\boldsymbol{x}'[u, \boldsymbol{q}_\alpha(t), t] = \frac{\partial \boldsymbol{x}'}{\partial t} + \frac{\partial \boldsymbol{x}'}{\partial q_{\alpha i}}\dot{q}_{\alpha i} \tag{7.32}$$

the velocities of corresponding points on the two curves are, in general, different, and so the coincidence is only instantaneous. According to Equation (7.32), the motion of C_α is due partly to the underlying motion of $C_{\boldsymbol{q}_\alpha}$ and partly to the motion of particle α. Now $\bar{L}_1$ may be written in terms of line integrals over the curves C_α. Thus

$$\bar{L}_1 = \frac{Q}{c}\dot{\boldsymbol{R}}\,.\,\boldsymbol{a}(\boldsymbol{R}, t) - Q\phi(\boldsymbol{R}, t) + \sum_\alpha e_\alpha \int_{C_\alpha} \boldsymbol{e}(\boldsymbol{x}', t)\,.\,\mathrm{d}\boldsymbol{x}'$$
$$+ \sum_\alpha \frac{e_\alpha}{c}\int_{C_\alpha} \dot{\boldsymbol{x}}'\,.\,\boldsymbol{b}(\boldsymbol{x}', t) \times \mathrm{d}\boldsymbol{x}' \tag{7.33}$$

The first two terms correspond to the Lagrangian for a collective particle of charge Q located at $\boldsymbol{R}$ and interacting with the field described by the Coulomb potentials $\boldsymbol{a}$ and ϕ. The second two terms may be written as integrals over polarization and magnetization fields interacting with the electric and magnetic induction fields. For if $\boldsymbol{p}$ and $\boldsymbol{m}$ are defined by Equations (7.5) and (7.6), then $\bar{L}_1$ becomes

$$\bar{L}_1 = \iiint \left(\frac{1}{c}\boldsymbol{j}_{\text{true}}\,.\,\boldsymbol{a} - \rho_{\text{true}}\phi + \boldsymbol{p}\,.\,\boldsymbol{e} + \boldsymbol{m}\,.\,\boldsymbol{b}\right)\mathrm{d}^3x \tag{7.34}$$

with the true charge and current densities being those of the total charge Q when associated with the reference point $\boldsymbol{R}$. For a neutral system ($Q = 0$), the true charge and current densities disappear and the line integral Lagrangian $\bar{L}$ involves only the coupling between the polarization and magnetization and the electric and magnetic induction fields. Whereas previously the motion of the curves C_α that define the line-integral polarization and magnetization fields was not specified, beyond the fact that C_α must start at $\boldsymbol{R}$ and end at q_α, this motion is now determined by the nature of the curves $C_{\boldsymbol{x}}$ that define the line integral vector and scalar potentials and by the motion of the charged particles in the electromagnetic field.

The equality of the expressions (7.27) and (7.34) for $\bar{L}_1$ can be demonstrated in another way (Power 1964, Wooley and Cordle 1973). If the gauge function χ has the form given in Equation (2.26), $\boldsymbol{a}$ being the vector potential in the Coulomb gauge, then

$$\iiint \rho\chi \, d^3x = \sum_\alpha e_\alpha \int_{C_\alpha} \boldsymbol{a}(\boldsymbol{x}', t) \cdot d\boldsymbol{x}'$$
$$= \iiint \boldsymbol{p} \cdot \boldsymbol{a} \, d^3x \tag{7.35}$$

where $\boldsymbol{p}$ is the line integral polarization field defined in Equation (7.5). It follows from Equation (7.29) that

$$\bar{L}_1 = \iiint \left(\frac{1}{c} \boldsymbol{j} \cdot \bar{\boldsymbol{a}} - \rho \bar{\phi} \right) d^3x$$
$$= \iiint \left(\frac{1}{c} \boldsymbol{j} \cdot \boldsymbol{a} - \rho \phi \right) d^3x - \frac{1}{c} \frac{d}{dt} \iiint \boldsymbol{p} \cdot \boldsymbol{a} \, d^3x \tag{7.36}$$

Instead of substituting, as above, the expressions (2.29) and (2.30) for $\bar{\boldsymbol{a}}$ and $\bar{\phi}$ in terms of $\boldsymbol{e}$ and $\boldsymbol{b}$, we can substitute the expressions (7.3) and (7.4) for ρ and $\boldsymbol{j}$ in terms of $\boldsymbol{p}$ and $\boldsymbol{m}$. This gives

$$\bar{L}_1 = \iiint \left[\frac{1}{c} \boldsymbol{j}_{\text{true}} \cdot \boldsymbol{a} - \rho_{\text{true}} \phi - \frac{1}{c} \boldsymbol{p} \cdot \dot{\boldsymbol{a}} + (\nabla \times \boldsymbol{m}) \cdot \boldsymbol{a} + (\nabla \cdot \boldsymbol{p}) \phi \right] d^3x \tag{7.37}$$

If the last two terms are integrated by parts and Equations (2.10) and (2.11) are used, then Equation (7.34) is immediately recovered.

It was seen in Section 2.4 that changing the curves $C_{\boldsymbol{x}}$ amounts to a gauge transformation of the line integral vector and scalar potentials, and it was seen in Section 7.2 that changing the curves C_α amounts to a "gauge" transformation of the line integral polarization and magnetization fields. The connection between these transformations will now be examined. Associated with the two sets of line integral potentials are two line integral Lagrangians $\bar{L}^{(1)}$ and $\bar{L}^{(2)}$ and these, since they are obtained from each other through gauge transformations of the potentials, are equivalent. Indeed we have, analogously to Equation (7.29),

$$\bar{L}^{(2)} = \bar{L}^{(1)} - \frac{1}{c} \frac{d}{dt} \iiint \rho \chi^{(12)} \, d^3x \tag{7.38}$$

with $\chi^{(12)}$ being given by Equation (2.32) or (2.33). $\bar{L}_1^{(1)}$ and $\bar{L}_1^{(2)}$ can be expressed, as in Equation (7.34), in terms of polarization and magnetization fields. The curves $C_{\boldsymbol{x}}^{(1)}$ determine curves $C_\alpha^{(1)}$ and thus polarization and magnetization fields $\boldsymbol{p}^{(1)}$ and $\boldsymbol{m}^{(1)}$, defined as in Equations (7.5) and (7.6). Similarly the curves $C_{\boldsymbol{x}}^{(2)}$ determine curves $C_\alpha^{(2)}$ and polarization and magnetization fields $\boldsymbol{p}^{(2)}$ and $\boldsymbol{m}^{(2)}$. The two sets of line integral polarization and magnetization fields satisfy equations like (7.3) and (7.4) with the same total charge and current densities and with the true charge and current densities centred on either $\boldsymbol{R}_1$ or $\boldsymbol{R}_2$, as the case may be. The two sets of polarization and magnetization fields must therefore be related to each other through the transformation (7.11) and (7.12) with $\boldsymbol{v}^{(12)}$ and $s^{(12)}$ being given by Equations (7.21) and (7.22). We may take the curve C_{12} that appears in this transformation to be the same curve that was used to express the gauge function $\chi^{(12)}$ in the form (2.33), and suppose that $\Sigma_{\boldsymbol{x}}$ varies smoothly with $\boldsymbol{x}$, although, as was indicated in Section 2.4, this assumption is not necessary so far as the gauge function itself is concerned. Σ_α can then be taken as that surface which, for each time t, coincides instantaneously with the surfaces $\Sigma_{\boldsymbol{q}_\alpha}$, just as the curve C_α coincides instantaneously with the curves $C_{\boldsymbol{q}_\alpha}$. The surfaces Σ_α, and hence the vector and scalar fields (7.21) and (7.22), are thus determined by the surfaces $\Sigma_{\boldsymbol{x}}$ and by the trajectories of the charged particles. In this way a gauge transformation of the potentials, caused by altering the curves $C_{\boldsymbol{x}}$, induces a "gauge" transformation of the polarization and magnetization fields, caused by altering the curves C_α.

PROBLEMS

(1) Show that if

$$\boldsymbol{p}^{(12)}(\boldsymbol{x}, t) = Q\int_{C_{12}} \mathrm{d}\boldsymbol{x}'\, \delta(\boldsymbol{x}-\boldsymbol{x}')$$

and if $\chi^{(12)}$ and $\boldsymbol{v}^{(12)}$ are defined by Equations (2.33) and (7.21), respectively, then

$$\iiint \rho\chi^{(12)}\, \mathrm{d}^3x = \iiint (\boldsymbol{v}^{(12)} \cdot \boldsymbol{b} - \boldsymbol{p}^{(12)} \cdot \boldsymbol{a})\, \mathrm{d}^3x$$

Use this to demonstrate the equivalence of the line integral Lagrangians $\bar{L}^{(1)}$ and $\bar{L}^{(2)}$ through the polarization and magnetization fields, instead of through the vector and scalar potentials.

(2) Suppose the potentials $\boldsymbol{a}$ and ϕ are both regarded as generalized coordinates in the Lagrangian L of Equation (7.23) and that no gauge condition is imposed on them or their variations in the application of Hamilton's principle. Show that the resulting Euler–Lagrange equations still lead to the Maxwell–Lorentz equations with sources but that the field conjugate to ϕ vanishes, so that the Lagrangian is degenerate.

7.4 The multipolar Hamiltonian

The addition to the classical Lagrangian of the total time derivative of a function of the generalized coordinates leads to a canonical transformation that changes the canonical momenta but not the coordinates. For the corresponding transformation in quantum theory, the operators representing the new canonical variables are related to those representing the old by a unitary transformation. To see this, let us consider again a system with n degrees of freedom and subtract from the Lagrangian the total time derivative of a real function $W(q)$, so that the new momenta are given by $\bar{p}_i = p_i - \partial W/\partial q_i$. The relation between the quantum mechanical operators is then $\bar{p}_i = U p_i U^{-1}$ with $U = \exp(iW/\hbar)$, as may be shown by using Schrödinger's representation or the operator identity of Theorem 1, Appendix D. (In this case the operator series terminates because $[W, p_i]$ is a function of q and hence commutes with W.) Also, since q_i commutes with W, $q_i = U q_i U^{-1}$. As W is a real function, $W(q)$ is a Hermitian operator and U is unitary. This guarantees that the new canonical variables, just like the old, are represented by Hermitian operators that satisfy the canonical commutation relations.

The multipolar form of the Hamiltonian in non-relativistic quantum electrodynamics is related to the minimal-coupling form through a canonical transformation induced by the unitary operator $\exp(iW/\hbar)$, or U, in which the generator W, which is a functional of the transverse vector potential $\boldsymbol{a}$ and a function of the particle coordinates q, is given by

$$W[\boldsymbol{a}; q) = \frac{1}{c}\iiint \boldsymbol{p}\,.\,\boldsymbol{a}\, \mathrm{d}^3x = \frac{1}{c}\iiint \rho\chi\, \mathrm{d}^3x \qquad (7.39)$$

We shall not carry out the transformation for the most general choice of the curves $C_{\boldsymbol{x}}$ that define the gauge function χ, or of the curves C_α that define the polarization field $\boldsymbol{p}$. The analysis will be restricted to the case in which (i) the fixed-nuclei approximation is used, so that the coordinates and momenta of the electrons only are treated as dynamical variables and ρ and $\boldsymbol{j}$ refer to the electronic charge and current densities alone, (ii) the reference point is fixed in space (typically, within the nuclear framework, if a multipolar expansion is to be employed) and is thus a c-number rather than a function of the operators representing the dynamical variables, and (iii) the curves $C_{\boldsymbol{x}}$ are specified, as in Section 2.3, by functions of the form $\boldsymbol{x}'(u, \boldsymbol{x})$, $u_1 \leqslant u \leqslant u_2$, with u_1 corresponding to $\boldsymbol{R}$ and u_2 to $\boldsymbol{x}$. These assumptions enable the essential features of the transformation to be illustrated in a relatively simple way. On the curves C_α, $\boldsymbol{x}'$ is a definite function $\boldsymbol{x}'(u, \boldsymbol{q}_\alpha)$ of u and $\boldsymbol{q}_\alpha$ and is consequently a q-number dynamical variable. For example, in the case of straight-line paths we have $\boldsymbol{x}'=\boldsymbol{R}+u(\boldsymbol{q}_\alpha-\boldsymbol{R})$, $0\leqslant u\leqslant 1$. The time derivative of a point on the curve C_α is determined dynamically (rather than kinematically) through the relation

$$\dot{\boldsymbol{x}}'=\frac{1}{i\hbar}[\boldsymbol{x}', H]=\frac{1}{2}\left\{\frac{\partial \boldsymbol{x}'}{\partial q_{\alpha i}}\dot{q}_{\alpha i}+\dot{q}_{\alpha i}\frac{\partial \boldsymbol{x}'}{\partial q_{\alpha i}}\right\} \tag{7.40}$$

in which the order of the operators $\dot{q}_{\alpha i}$ and $\partial\boldsymbol{x}'/\partial q_{\alpha i}$ is, in general, significant. We shall work in the Schrödinger picture of the motion, so that the gauge function χ, the polarization field $\boldsymbol{p}$, etc., are functions of the field point $\boldsymbol{x}$ but not of the time t.

The canonical transformation is performed by introducing into the Hamiltonian new canonical momenta defined by

$$\bar{\boldsymbol{p}}_\alpha=U\boldsymbol{p}_\alpha U^{-1} \qquad \bar{\boldsymbol{\pi}}(\boldsymbol{x})=U\boldsymbol{\pi}(\boldsymbol{x})U^{-1} \tag{7.41}$$

Using the second form of the generator W given by Equation (7.39) we obtain for the transformed particle momenta

$$\begin{aligned}\bar{p}_{\alpha i}&=p_{\alpha i}-\frac{\partial W}{\partial q_{\alpha i}}=p_{\alpha i}+\frac{e}{c}\frac{\partial\chi}{\partial q_{\alpha i}}\\&=p_{\alpha i}+\frac{e}{c}a_i(\boldsymbol{q}_\alpha)-\frac{e}{c}\bar{a}_i(\boldsymbol{q}_\alpha)\end{aligned} \tag{7.42}$$

where $\bar{\boldsymbol{a}}$ is the line integral potential. It should be noted that, because of the assumptions made about the curves $C_{\boldsymbol{x}}$, $\partial\chi/\partial q_{\alpha i}=[\partial\chi/\partial x_i]_{\boldsymbol{x}=\boldsymbol{q}_\alpha}$. This would not be true if, e.g., the reference point were a function of the $\boldsymbol{q}_\alpha$ instead of being fixed. For the transformed field momentum we obtain, using now the first form of the generator W given by Equation (7.39),

$$\begin{aligned}\bar{\pi}_i(\boldsymbol{x})&=\pi_i(\boldsymbol{x})-\frac{\delta W}{\delta a_i^\perp(\boldsymbol{x})}=-\frac{1}{4\pi c}e_i^\perp(\boldsymbol{x})-\frac{1}{c}p_i^\perp(\boldsymbol{x})\\&=-\frac{1}{4\pi c}d_i^\perp(\boldsymbol{x})\end{aligned} \tag{7.43}$$

$\boldsymbol{d}$ being the microscopic displacement vector corresponding to the particular polarization field $\boldsymbol{p}$ that is associated with the curves C_α. Since the coordinates q and $\boldsymbol{a}$ are invariant under the transformation, the Hamiltonian may be written in terms of the new variables as

$$\begin{aligned}H=&\frac{1}{8\pi}\iiint\{(4\pi c\bar{\boldsymbol{\pi}})^2+(\nabla\times\boldsymbol{a})^2\}\,\mathrm{d}^3x+4\pi c\iiint\boldsymbol{p}^\perp\cdot\bar{\boldsymbol{\pi}}\,\mathrm{d}^3x\\&+2\pi\iiint\boldsymbol{p}^{\perp 2}\,\mathrm{d}^3x+\frac{1}{2m}\sum_\alpha\left\{\bar{\boldsymbol{p}}+\frac{e}{c}\bar{\boldsymbol{a}}(\boldsymbol{q}_\alpha)\right\}^2+V(q)\end{aligned} \tag{7.44}$$

where α labels electrons only and V represents the potential of the external electrostatic fields, including that of the fixed nuclei. (We assume there are no external magnetostatic fields.) The fourth term in Equation (7.44) is, as some straightforward manipulation shows, the sum of a "kinetic energy" term $\sum_\alpha \bar{\boldsymbol{p}}_\alpha^2/(2m)$ and an "interaction energy" term

$$-\iiint \tilde{\boldsymbol{m}} \cdot \boldsymbol{b} \, \mathrm{d}^3x + \frac{1}{2}\iiint \mathrm{d}^3x \iiint \mathrm{d}^3y \, o_{ij}(\boldsymbol{x}, \boldsymbol{y}) b_i(\boldsymbol{x}) b_j(\boldsymbol{y})$$

in which the magnetization field $\tilde{\boldsymbol{m}}$ is defined by

$$\tilde{\boldsymbol{m}}(\boldsymbol{x}) = -\frac{e}{2mc}\sum_\alpha \int_{C_\alpha} \left\{ \mathrm{d}\boldsymbol{x}' \, \delta(\boldsymbol{x}-\boldsymbol{x}') \times \frac{\partial \boldsymbol{x}'}{\partial q_{\alpha k}} \bar{p}_{\alpha k} - \bar{p}_{\alpha k} \frac{\partial \boldsymbol{x}'}{\partial q_{\alpha k}} \times \delta(\boldsymbol{x}-\boldsymbol{x}') \, \mathrm{d}\boldsymbol{x}' \right\} \tag{7.45}$$

and the diamagnetization field $o_{ij}(\boldsymbol{x}, \boldsymbol{y})$ by

$$o_{ij}(\boldsymbol{x}, \boldsymbol{y}) = \frac{e^2}{mc^2}\sum_\alpha \int_{C_\alpha} \left(\mathrm{d}\boldsymbol{x}' \times \frac{\partial \boldsymbol{x}'}{\partial q_{\alpha k}} \right)_i \delta(\boldsymbol{x}-\boldsymbol{x}') \int_{C_\alpha} \left(\mathrm{d}\boldsymbol{y}' \times \frac{\partial \boldsymbol{y}'}{\partial q_{\alpha k}} \right)_j \delta(\boldsymbol{y}-\boldsymbol{y}') \tag{7.46}$$

It follows that the multipolar Hamiltonian, when expressed in terms of the electromagnetic fields rather than the vector potential and its conjugate momentum, becomes

$$H = \bar{H}_0 + \bar{H}_{\text{int}} \equiv \bar{H}_{\text{rad}} + \bar{H}_{\text{atom}} + \bar{H}_{\text{int}} \tag{7.47}$$

where

$$\bar{H}_{\text{rad}} = \frac{1}{8\pi}\iiint (\boldsymbol{d}^{\perp 2} + \boldsymbol{b}^2) \, \mathrm{d}^3x \tag{7.48}$$

$$\bar{H}_{\text{atom}} = \frac{1}{2m}\sum_\alpha \bar{\boldsymbol{p}}_\alpha^2 + V(q) \tag{7.49}$$

and

$$\begin{aligned} \bar{H}_{\text{int}} = & -\iiint \boldsymbol{p} \cdot \boldsymbol{d}^\perp \, \mathrm{d}^3x - \iiint \tilde{\boldsymbol{m}} \cdot \boldsymbol{b} \, \mathrm{d}^3x \\ & + \frac{1}{2}\iiint \mathrm{d}^3x \iiint \mathrm{d}^3y \, o_{ij}(\boldsymbol{x}, \boldsymbol{y}) b_i(\boldsymbol{x}) b_j(\boldsymbol{y}) \\ & + 2\pi \iiint \boldsymbol{p}^{\perp 2} \, \mathrm{d}^3x \end{aligned} \tag{7.50}$$

In the new interaction Hamiltonian $\bar{H}_{\text{int}}$, the polarization, magnetization and diamagnetization fields are coupled directly to the transverse electric displacement vector and the magnetic induction field. The transverse vector potential, though still a canonical coordinate, does not appear explicitly. The term proportional to the integral of the square of the transverse polarization field in Equation (7.50) is a self-energy term which does not contribute to any process involving a change of state of the electromagnetic field. Its effect is nevertheless important when the multipolar Hamiltonian is used in carrying out the non-relativistic mass-renormalization programme (Power and Zienau 1959). The order in which the non-commuting operators occur in the expression (7.45) for the magnetization field $\tilde{\boldsymbol{m}}$ is prescribed by the unitary transformation and does not have to be postulated separately so as to make $\tilde{\boldsymbol{m}}$ Hermitian, as would be the case if the canonical transformation were first carried out on the classical Hamiltonian and canonical quantization then applied. (In the minimal-coupling form of the Hamiltonian, it will be remembered, the order of the operators $\boldsymbol{p}_\alpha$ and $\boldsymbol{a}(\boldsymbol{q}_\alpha)$ is immaterial, due to the transverse nature of $\boldsymbol{a}$.) The relation of the magnetization field $\tilde{\boldsymbol{m}}$ to the quantum mechanical version of the magnetization field $\boldsymbol{m}$ should be made clear. The former is defined in terms of the new canonical momenta and the latter in terms of the velocities. We have, from Equations (2.12) and (7.40),

$$\begin{aligned}
\tilde{m}_i(\boldsymbol{x}) - \iiint o_{ij}(\boldsymbol{x}, \boldsymbol{y}) b_j(\boldsymbol{y})\, \mathrm{d}^3 y \\
&= -\frac{e}{2c} \sum_\alpha \int_{C_\alpha} \left\{ \mathrm{d}\boldsymbol{x}'\, \delta(\boldsymbol{x}-\boldsymbol{x}') \times \frac{\partial \boldsymbol{x}'}{\partial q_{\alpha k}} \dot{q}_{\alpha k} - \dot{q}_{\alpha k} \frac{\partial \boldsymbol{x}'}{\partial q_{\alpha k}} \times \delta(\boldsymbol{x}-\boldsymbol{x}')\, \mathrm{d}\boldsymbol{x}' \right\}_i \\
&= -\frac{e}{2c} \sum_\alpha \int_{C_\alpha} \left\{ \mathrm{d}\boldsymbol{x}'\, \delta(\boldsymbol{x}-\boldsymbol{x}') \times \dot{\boldsymbol{x}}' - \dot{\boldsymbol{x}}' \times \delta(\boldsymbol{x}-\boldsymbol{x}')\, \mathrm{d}\boldsymbol{x}' \right\}_i \\
&\equiv m_i(\boldsymbol{x})
\end{aligned} \tag{7.51}$$

Note, however, that the Hamiltonian density for the coupling of the magnetic induction field to the charges is not simply $-\boldsymbol{m} \cdot \boldsymbol{b}$, since the part of $\boldsymbol{m}$ that depends explictly on $\boldsymbol{b}$ in Equation (7.51) appears in the Hamiltonian with a factor of 1/2. Note also that it is $\boldsymbol{m}$ (which has a more immediate physical significance) and not $\tilde{\boldsymbol{m}}$ that gives rise to the magnetization current density as in Equation (7.4), the quantum mechanical operator for the total current density $\boldsymbol{j}$ given by Equation (4.27) being symmetrized in the same way as that for $\boldsymbol{m}$.

The unitary operator U has been interpreted as effecting a change from one set of canonical variables to another. This agrees with the interpretation usually given to canonical transformations in classical mechanics. Neither the Hamiltonian operator H nor the correspondence between rays and physical states has been altered—the same Hamiltonian is merely a new functional or function of new canonical variables. In its minimal-coupling form $H = F[\boldsymbol{\pi}, \boldsymbol{a}; p, q)$, i.e. is a functional of $\boldsymbol{\pi}$ and $\boldsymbol{a}$ and a function of p and q, and in its multipolar form $H = \bar{F}[\bar{\boldsymbol{\pi}}, \boldsymbol{a}; \bar{p}, q)$, i.e. is a functional of $\bar{\boldsymbol{\pi}}$ and $\boldsymbol{a}$ and a function of $\bar{p}$ and q, where $\bar{\boldsymbol{\pi}} = U\boldsymbol{\pi}U^{-1}$ and $\bar{p} = UpU^{-1}$. There are many other ways in which unitary operators may be used in quantum mechanics. We have seen in Chapter 5 that the change in the description of a physical system on going from one observer to another may be achieved by a unitary linear or antilinear operator acting, by convention, on the state vectors only. Unitary operators are also used to change the correspondence between states and rays and between dynamical variables and operators. Thus if U is a unitary operator, we may associate the vector $|S'\rangle = U|S\rangle$ with that state which was formerly associated with the vector $|S\rangle$ and simultaneously associate the Hermitian operator $\Omega' = U\Omega U^{-1}$ with that dynamical variable which was formerly associated with the Hermitian operator Ω. These two representations are isomorphic and no difference can arise from choosing one rather than the other, since the physical predictions of the theory are based on the relations between the operators and the state vectors and these are the same in each case. Thus Ω and Ω' have the same eigenvalues, $\langle R'|\,\Omega'\,|S'\rangle = \langle R|\,\Omega\,|S\rangle$, etc. We now have a new Hamiltonian *operator* given by

$$\begin{aligned} H' = UHU^{-1} &= UF[\boldsymbol{\pi}, \boldsymbol{a}, p, q]U^{-1} \\ &= F[\boldsymbol{\pi}', \boldsymbol{a}'; p', q') \end{aligned} \tag{7.52}$$

but its *functional form* is unchanged, the new Hamiltonian being the same functional or function of the new operators as the old Hamiltonian is of the old operators. We emphasize that in this section we have dealt with (i) a fixed observer and (ii) a fixed correspondence between states and rays and between dynamical variables and operators.

The foregoing treatment can be generalized in two ways. First, motion of the curves $C_{\boldsymbol{x}}$ and of the reference point $\boldsymbol{R}$ can be taken into account, whether or not the nuclei are regarded as fixed. Terms depending on this motion then appear in the Hamiltonian.† It is sometimes convenient to define magnetization fields and magnetic multipole moments in terms of velocities or momenta relative to the reference point, if this is moving, instead of relative to an inertial coordinate system. The total magnetization current then contains a contribution—the so-called Röntgen current—that depends explicitly on the motion of $\boldsymbol{R}$ (de Groot 1969,

† For the case in which the integration paths are straight lines but $\boldsymbol{R}$ is taken to be the centre of mass of the total aggregate (including the nuclei) and may thus have an arbitrary non-relativistic motion, see Felderhof and Adu-Gyamfi (1974) and Healy (1977a).

Power and Thirunamachandran 1971, Babiker *et al.* 1973, Healy 1977b). A second generalization can be made by applying the formalism to an assemblage of several non-overlapping aggregates (atoms, molecules, etc.). For each aggregate ξ a reference point $\boldsymbol{R}^{(\xi)}$ may be chosen and polarization fields $\boldsymbol{p}^{(\xi)}$ defined as before. The total field $\boldsymbol{p}$ is then the sum over all the aggregates of the partial fields $\boldsymbol{p}^{(\xi)}$. The Hamiltonian for the complete system now contains instantaneous pairwise interaggregate Coulomb energies as well as the intra-aggregate Coulomb energies that are included in H_0 or $\bar{H}_0$. However, the multipolar Hamiltonian also contains the term $2\pi \iiint \boldsymbol{p}^{\perp 2}\, \mathrm{d}^3x$ which corresponds to the self-energy term in Equation (7.50) for a single aggregate but which can now be decomposed into diagonal and off-diagonal parts:

$$2\pi\iiint \boldsymbol{p}^{\perp 2}\,\mathrm{d}^3x = 2\pi\sum_{\xi}\iiint \boldsymbol{p}^{(\xi)\perp 2}\,\mathrm{d}^3x + 2\pi\sum\sum_{\xi\neq\eta}\iiint \boldsymbol{p}^{(\xi)\perp}\cdot\boldsymbol{p}^{(\eta)\perp}\,\mathrm{d}^3x \tag{7.53}$$

For neutral aggregates, for which the true charge densities vanish, the longitudinal polarization fields $\boldsymbol{p}^{(\xi)\|}$ are given, according to Equations (7.5) and (C.5), by

$$\boldsymbol{p}^{(\xi)\|}(\boldsymbol{x}) = \frac{1}{4\pi}\nabla\sum_{\alpha(\xi)}\frac{e_{\alpha(\xi)}}{|\boldsymbol{x}-\boldsymbol{q}_{\alpha(\xi)}|} \tag{7.54}$$

where the sum is to extend over electrons and nuclei alike, even if the latter are regarded as fixed. The relation (7.54) may be used (Power and Zienau 1959, Woolley 1971a, b) to eliminate from the Hamiltonian the interaggregate Coulomb energy. We have first

$$\frac{1}{2}\sum\sum_{\xi\neq\eta}\sum_{\alpha(\xi)}\sum_{\beta(\eta)}\frac{e_{\alpha(\xi)}e_{\beta(\eta)}}{|\boldsymbol{q}_{\alpha(\xi)}-\boldsymbol{q}_{\beta(\eta)}|} = 2\pi\sum\sum_{\xi\neq\eta}\iiint \boldsymbol{p}^{(\xi)\|}\cdot\boldsymbol{p}^{(\eta)\|}\,\mathrm{d}^3x \tag{7.55}$$

as follows from Equation (7.54) and integration by parts. Addition of the right-hand sides of Equations (7.53) and (7.55) then yields the expression

$$2\pi\sum_{\xi}\iiint \boldsymbol{p}^{(\xi)\perp 2}\,\mathrm{d}^3x + 2\pi\sum\sum_{\xi\neq\eta}\iiint \boldsymbol{p}^{(\xi)}\cdot\boldsymbol{p}^{(\eta)}\,\mathrm{d}^3x$$

because the integral of the scalar product of a longitudinal and a transverse vector field vanishes, if the fields drop off sufficiently rapidly at infinity. The first part of this expression is a sum over self-energies of the individual aggregates. The second part is a contact term which can be neglected unless different aggregates overlap significantly, since, in contrast to their transverse or longitudinal parts, the polarization fields $\boldsymbol{p}^{(\xi)}$ are of a local character. Thus, so long as the aggregates are electrically neutral and well separated, the interaggregate Coulomb energies are effectively cancelled by the off-diagonal terms in Equation (7.53).

PROBLEMS

(1) Show that if the integration paths are chosen to be straight lines, the multipolar interaction Hamiltonian, with the self-energy term being omitted, has the expansion

$$\begin{aligned}\bar{H}_{\text{int}} = &-\sum_{r=1}^{\infty} p^{(r)}_{i_1 i_2 \ldots i_r} \nabla_{i_2} \ldots \nabla_{i_r} d^{\perp}_{i_1}(\boldsymbol{R})\\ &-\sum_{r=1}^{\infty} m^{(r)}_{i_1 i_2 \ldots i_r} \nabla_{i_2} \ldots \nabla_{i_r} b_{i_1}(\boldsymbol{R})\\ &+\frac{1}{2}\sum_{r=1}^{\infty}\sum_{s=1}^{\infty} b_{i_1}(\boldsymbol{R}) \overleftarrow{\nabla}_{i_2} \ldots \overleftarrow{\nabla}_{i_r} o^{(r,s)}_{i_1 i_2 \ldots i_r; j_1 j_2 \ldots j_s} \overrightarrow{\nabla}_{j_2} \ldots \overrightarrow{\nabla}_{j_s} b_{j_1}(\boldsymbol{R})\end{aligned}$$

where the electric, magnetic and diamagnetic multipole moment operators are defined by

$$p^{(r)}_{i_1 i_2 \ldots i_r} = -\frac{e}{r!}\sum_{\alpha} \rho_{\alpha i_1} \rho_{\alpha i_2} \ldots \rho_{\alpha i_r}$$

$$m^{(r)}_{i_1 i_2 \ldots i_r} = -\frac{r}{(r+1)!}\frac{e}{2mc}\sum_{\alpha} (\bar{l}_{\alpha i_1} \rho_{\alpha i_2} \ldots \rho_{\alpha i_r} + \rho_{\alpha i_2} \ldots \rho_{\alpha i_r} \bar{l}_{\alpha i_1})$$

and

$$\begin{aligned}o^{(r,s)}_{i_1 i_2 \ldots i_r; j_1 j_2 \ldots j_s} = &\frac{rs}{(r+1)!\,(s+1)!}\frac{e^2}{mc^2}\sum_{\alpha} (\rho_{\alpha}^2 \delta_{i_1 j_1}\\ &- \rho_{\alpha i_1} \rho_{\alpha j_1}) \rho_{\alpha i_2} \ldots \rho_{\alpha i_r} \rho_{\alpha j_2} \ldots \rho_{\alpha j_s}\end{aligned}$$

respectively, and where $\boldsymbol{\rho}_{\alpha}$ is the position operator $\boldsymbol{q}_{\alpha} - \boldsymbol{R}$ and $\bar{\boldsymbol{l}}_{\alpha}$ the angular momentum operator $\boldsymbol{\rho}_{\alpha} \times \bar{\boldsymbol{p}}_{\alpha}$ of electron α relative to $\boldsymbol{R}$. The notation used for the partial derivatives means that

$$f(\boldsymbol{R})\overleftarrow{\nabla} = \overrightarrow{\nabla} f(\boldsymbol{R}) = \nabla f(\boldsymbol{R}) = [\nabla f(\boldsymbol{x})]_{\boldsymbol{x}=\boldsymbol{R}}$$

(2) Show that changing the curves C_{α} in the multipolar form of the Hamiltonian is equivalent to carrying out a canonical transformation with generating function

$$\begin{aligned}W^{(12)}[\boldsymbol{a}; q) &= \frac{1}{c}\iiint [\boldsymbol{v}^{(12)} \cdot \boldsymbol{b} - \boldsymbol{p}^{(12)} \cdot \boldsymbol{a}]\, \mathrm{d}^3 x\\ &= \frac{1}{c}\iiint \rho \chi^{(12)}\, \mathrm{d}^3 x\end{aligned}$$

where the notation is as in Problem 1, Section 7.3.

(3) Show that if an external static, though not necessarily uniform, magnetic induction field $\boldsymbol{B}$ is present, the multipolar interaction Hamiltonian should be modified by replacing $\boldsymbol{b}$ by $\boldsymbol{b}+\boldsymbol{B}$.

7.5 Applications

ATOMIC FIELD EQUATIONS

It is interesting to derive the Heisenberg equations of motion for some of the variables that appear in the multipolar form of the Hamiltonian (see also Babiker *et al.* 1974, Babiker 1975, Healy 1977a, Power and Thirunamachandran 1978). The new canonical coordinates and momenta obey the same commutation rules as the old, since they are related to them through a unitary transformation. Thus

$$[q_{\alpha i}, \bar{p}_{\beta j}] = i\hbar\, \delta_{\alpha\beta}\, \delta_{ij} \qquad [a_i(\boldsymbol{x}), \bar{\pi}_j(\boldsymbol{x}')] = i\hbar\, \delta^{\perp}_{ij}(\boldsymbol{x}-\boldsymbol{x}') \tag{7.56}$$

with all other canonical commutators being zero. The Heisenberg equation for the vector potential $\boldsymbol{a}$ merely confirms that the new field momentum $\bar{\boldsymbol{\pi}}$ is essentially the transverse displacement vector $\boldsymbol{d}^{\perp}$, in contrast to the old momentum $\boldsymbol{\pi}$ which is, apart from a factor, the transverse electric field $\boldsymbol{e}^{\perp}$. From the canonical commutation relations we obtain

$$\begin{aligned}\dot{a}_i(\boldsymbol{x}) &= \frac{1}{i\hbar}[a_i(\boldsymbol{x}), H]\\ &= 4\pi c^2 \bar{\pi}_j(\boldsymbol{x}) + 4\pi c p_i^{\perp}(\boldsymbol{x})\end{aligned} \tag{7.57}$$

since $\bar{\boldsymbol{\pi}}$ is transverse. Hence

$$\bar{\boldsymbol{\pi}} = -\frac{1}{4\pi c}(\boldsymbol{e}^{\perp} + 4\pi \boldsymbol{p}^{\perp}) = -\frac{1}{4\pi c}\boldsymbol{d}^{\perp} \tag{7.58}$$

in agreement with Equation (7.43). The Heisenberg equation for $\bar{\boldsymbol{\pi}}$ leads to the atomic field equation that connects $\boldsymbol{d}^{\perp}$ to the microscopic magnetic field $\boldsymbol{h}$. In view of the relation (7.51), we have

$$\begin{aligned}\dot{\bar{\pi}}_i(\boldsymbol{x}) &= \frac{1}{i\hbar}[\bar{\pi}_i(\boldsymbol{x}), H]\\ &= \left[-\frac{1}{4\pi}\nabla\times\boldsymbol{b}(\boldsymbol{x}) + \nabla\times\boldsymbol{m}(\boldsymbol{x})\right]_i\end{aligned} \tag{7.59}$$

According to Equation (7.58), this may be written as

$$\nabla\times\boldsymbol{h} = \frac{1}{c}\dot{\boldsymbol{d}}^{\perp} \tag{7.60}$$

where, as before, $\boldsymbol{h}=\boldsymbol{b}-4\pi\boldsymbol{m}$. Equation (7.60) agrees with the atomic field Equation (7.2), since for a fixed reference point the true current density must vanish. This does not require the true charge density to vanish, although it must be constant in time. If the reference point is not fixed and the dynamical part of the aggregate of charges is not electrically neutral (and this happens when (i) only electrons are treated as dynamical entities or (ii) nuclei are treated as part of the dynamical system but together with the electrons form a positive or negative ion), then the transverse vector potential, evaluated at the reference point, is coupled to the true current density in the Lagrangian and appears explicitly in the multipolar Hamiltonian.

PERTURBATION THEORY

Equations (6.21) and (7.47) represent different partitionings of the same Hamiltonian operator H. Since $\bar{H}_{\text{rad}}=UH_{\text{rad}}U^{-1}$ and $\bar{H}_{\text{atom}}=UH_{\text{atom}}U^{-1}$, we have $\bar{H}_0=UH_0U^{-1}$. This relation does not hold between $\bar{H}_{\text{int}}$ and H_{int}, however, as the total Hamiltonian operator is not invariant under the unitary transformation. If new creation and annihilation operators are defined through the equations (of which either one implies the other)

$$\bar{a}^{(\lambda)\dagger}(\boldsymbol{k})=Ua^{(\lambda)\dagger}(\boldsymbol{k})U^{-1} \qquad \bar{a}^{(\lambda)}(\boldsymbol{k})=Ua^{(\lambda)}(\boldsymbol{k})U^{-1} \tag{7.61}$$

then $\boldsymbol{d}^{\perp}$ has the same expansion in terms of the new operators as $\boldsymbol{e}^{\perp}$ has in terms of the old, and a similar statement holds for $\bar{H}_{\text{rad}}$ and H_{rad}. This is true even when normal ordering relative to the new creation and annihilation operators is carried out on $\bar{H}_{\text{rad}}$, just as normal ordering relative to the old operators was carried out on H_{rad}. (Note that the infinite zero-point energy is the same in each case.) The vector potential $\boldsymbol{a}$, since it is invariant under the transformation, has the same expansion in terms of either set of operators. The new operators satisfy the same Bose commutation relations as the old. There is a one-to-one correspondence between the eigenstates of H_0 and those of $\bar{H}_0$, the vectors associated with these states being mapped into each other by U. Thus, e.g.,

$$|\bar{\boldsymbol{k}}_1, \bar{\lambda}_1; \ldots ; \bar{\boldsymbol{k}}_n, \bar{\lambda}_n; \bar{r}\rangle = U\,|\boldsymbol{k}_1, \lambda_1; \ldots ; \boldsymbol{k}_n, \lambda_n; r\rangle \tag{7.62}$$

where the vector on the left-hand side is an eigenvector of $\bar{H}_{\text{rad}}$ with eigenvalue $\hbar c(k_1+\ldots+k_n)$, an eigenvector of

$$\frac{1}{4\pi c}\iiint :\boldsymbol{d}^{\perp}\times\boldsymbol{b}:\,\mathrm{d}^3x$$

(i.e. of $\bar{\boldsymbol{P}}_{\text{rad}}$ rather than of $\boldsymbol{P}_{\text{rad}}$) with eigenvalue $\hbar(\boldsymbol{k}_1+\ldots+\boldsymbol{k}_n)$ and an eigenvector of $\bar{H}_{\text{atom}}$ with eigenvalue E_r.

It follows from the above that when the multipolar Hamiltonian partitioned as in Equation (7.47) is used in time-dependent perturbation theory, the resulting transition rates or cross-sections refer to processes $\bar{i} \to \bar{f}$ which differ, in general, from the corresponding processes $i \to f$ that arise when the minimal-coupling Hamiltonian, partitioned in the usual way, is used. Despite this, however, the on-energy-shell matrix elements for these processes are equal up to order e^2 (Healy and Woolley 1978), as we shall now demonstrate. The normalized eigenvectors, or approximate eigenvectors, for the initial and final states are $|\bar{i}\rangle$ and $|\bar{f}\rangle$ and $|i\rangle$ and $|f\rangle$, where $|\bar{i}\rangle = U\,|i\rangle$ and $|\bar{f}\rangle = U\,|f\rangle$, and the common eigenvalues are ε_i and ε_f, respectively. The matrix elements are given to order e^2 in the perturbation expansions by the expressions

$$M_{fi} = \langle f|\,H_{\text{int}}\,|i\rangle + \sum_{v} \frac{\langle f|\,H_{\text{int}}\,|v\rangle\langle v|\,H_{\text{int}}\,|i\rangle}{\varepsilon_i - \varepsilon_v} \tag{7.63}$$

and

$$\bar{M}_{fi} = \langle \bar{f}|\,\bar{H}_{\text{int}}\,|\bar{i}\rangle + \sum_{\bar{v}} \frac{\langle \bar{f}|\,\bar{H}_{\text{int}}\,|\bar{v}\rangle\langle \bar{v}|\,\bar{H}_{\text{int}}\,|\bar{i}\rangle}{\varepsilon_i - \varepsilon_v} \tag{7.64}$$

It is assumed that in the sums over the virtual states ε_i is not near any possible ε_v. We first express the matrix elements of $\bar{H}_{\text{int}}$ between any two eigenvectors $|\bar{a}\rangle$ and $|\bar{b}\rangle$ of $\bar{H}_0$ in terms of matrix elements between the corresponding eigenvectors $|a\rangle$ and $|b\rangle$ of H_0. Since $\bar{H}_{\text{int}} = H - \bar{H}_0$, we have

$$\langle \bar{a}|\,\bar{H}_{\text{int}}\,|\bar{b}\rangle = \langle a|\,(U^{-1}HU - H_0)\,|b\rangle = \langle a|\,\tilde{H}_{\text{int}}\,|b\rangle \tag{7.65}$$

where the operator $\tilde{H}_{\text{int}}$, which differs from both H_{int} and $\bar{H}_{\text{int}}$, is given by

$$\tilde{H}_{\text{int}} = U^{-1}HU - H_0 = U^{-1}H_0U - H_0 + U^{-1}H_{\text{int}}U \tag{7.66}$$

($\tilde{H}_{\text{int}}$ has the same functional form as $\bar{H}_{\text{int}}$, but is expressed in terms of $\boldsymbol{e}^{\perp}$ and p instead of $\boldsymbol{d}^{\perp}$ and $\bar{p}$.) Now from the canonical commutation relations and the operator identity of Theorem 1, Appendix D, we obtain

$$\begin{aligned} U^{-1}H_0U &= e^{-(i/\hbar)W}H_0e^{(i/\hbar)W} \\ &= H_0 - \left[\frac{i}{\hbar}\,W, H_0\right] + \frac{1}{2}\left[\frac{i}{\hbar}\,W, \left[\frac{i}{\hbar}\,W, H_0\right]\right] \end{aligned} \tag{7.67}$$

and

$$\begin{aligned} U^{-1}H_{\text{int}}U &= e^{-(i/\hbar)W}(eH_1 + e^2H_2)e^{(i/\hbar)W} \\ &= eH_1 - e\left[\frac{i}{\hbar}\,W, H_1\right] + e^2H_2 \end{aligned} \tag{7.68}$$

where

$$H_1 = \frac{1}{mc} \sum_{\alpha} \boldsymbol{p} \cdot \boldsymbol{a}(\boldsymbol{q}_\alpha) \tag{7.69}$$

and

$$H_2 = \frac{1}{2mc^2} \sum_{\alpha} \boldsymbol{a}^2(\boldsymbol{q}_\alpha) \tag{7.70}$$

The series (7.67) terminates because

$$[W, H_0] = -i\hbar c \iiint \boldsymbol{p} \cdot \boldsymbol{e}^{\perp} \, \mathrm{d}^3x \tag{7.71}$$

and

$$[W, [W, H_0]] = -4\pi(\hbar c)^2 \iiint \boldsymbol{p}^{\perp 2} \, \mathrm{d}^3x \tag{7.72}$$

so that $[W, [W, H_0]]$ commutes with W. Similarly the series (7.68) terminates because

$$[W, H_1] = \frac{i\hbar}{mc} \sum_{\alpha} \iiint \frac{\partial p_i(\boldsymbol{x})}{\partial q_{\alpha j}} a_i(\boldsymbol{x}) a_j(\boldsymbol{q}_\alpha) \, \mathrm{d}^3x \tag{7.73}$$

which also commutes with W. It follows from Equations (7.66)–(7.68) that

$$\tilde{H}_{\text{int}} = e\tilde{H}_1 + e^2\tilde{H}_2 \tag{7.74}$$

where

$$e\tilde{H}_1 = eH_1 - \left[\frac{i}{\hbar} W, H_0\right] \tag{7.75}$$

and is of order e, and where

$$e^2\tilde{H}_2 = e^2H_2 - e\left[\frac{i}{\hbar} W, H_1\right] + \frac{1}{2}\left[\frac{i}{\hbar} W, \left[\frac{i}{\hbar} W, H_0\right]\right] \tag{7.76}$$

and is of order e^2. Here we have used the fact that W is of order e while H_0, H_1 and H_2 are of order unity. The compound matrix elements can now be written as

$$M_{fi} = e\langle f| H_1 |i\rangle + e^2\langle f| H_2 |i\rangle + e^2 \sum_{v} \frac{\langle f| H_1 |v\rangle\langle v| H_1 |i\rangle}{\varepsilon_i - \varepsilon_v} \tag{7.77}$$

and

$$\bar{M}_{fi} = e\langle f| \tilde{H}_1 |i\rangle + e^2\langle f| \tilde{H}_2 |i\rangle + e^2 \sum_{v} \frac{\langle f| \tilde{H}_1 |v\rangle\langle v| \tilde{H}_1 |i\rangle}{\varepsilon_i - \varepsilon_v} \tag{7.78}$$

from which it is evident that they are both of order e^2. From the definitions (7.75) and (7.76) we then obtain, after some reduction making use of the completeness properties of the set of states v,

$$\bar{M}_{fi} = M_{fi} + (\varepsilon_f - \varepsilon_i)\left(\frac{i}{\hbar}\langle f|\, W\,|i\rangle + \frac{ie}{\hbar}\sum_v \frac{\langle f|\, W\,|v\rangle\langle v|\, H_1\,|i\rangle}{\varepsilon_i - \varepsilon_v} + \frac{1}{2\hbar^2}\langle f|\, W^2\,|i\rangle\right) \tag{7.79}$$

so that on the energy shell ($\varepsilon_f = \varepsilon_i$) we have $\bar{M}_{fi} = M_{fi}$, as was required to be shown.

Although the partitioning of the multipolar Hamiltonian into unperturbed and interaction parts differs from the corresponding partitioning of the minimal-coupling Hamiltonian and, moreover, depends on the choice of integration paths used initially to define the polarization field $\boldsymbol{p}$, we have just seen that the matrix elements for real first- or second-order processes are the same in each case and are independent of the paths. (Corresponding matrix elements for virtual processes are not equal, in general.) Thus the transition rates for emission and absorption (Einstein's coefficients) and the cross-section for Kramers–Heisenberg scattering may be obtained just as well from the multipolar as from the minimal-coupling Hamiltonian (Power 1964, Healy 1977c, Healy and Woolley 1978.) The terms in $\bar{H}_{\mathrm{int}}$ that are linear in $\boldsymbol{d}^{\perp}$ and $\boldsymbol{b}$ have non-vanishing first-order matrix elements, while the terms quadratic in $\boldsymbol{b}$ have non-vanishing second-order matrix elements. In electric dipole approximation, with $\boldsymbol{\mu} = -e\sum_\alpha (\boldsymbol{q}_\alpha - \boldsymbol{R})$, the multipolar interaction Hamiltonian is simply $-\boldsymbol{\mu} \cdot \boldsymbol{d}^{\perp}(\boldsymbol{R})$, whatever the choice of integration paths. Use of this interaction in second-order perturbation theory leads directly to the Kramers–Heisenberg formula, thus obviating the need for carrying out Dirac's transformation. Higher-order electric as well as magnetic and diamagnetic multipole moment contributions can readily be included, or the multipole series can be summed by using the exact expression for $\bar{H}_{\mathrm{int}}$. It should be noted that matrix elements of multipole moment operators taken between eigenvectors of $\bar{H}_{\mathrm{atom}}$, which occur when the multipolar Hamiltonian is used, can easily be expressed as matrix elements taken between eigenvectors of H_{atom}. Thus for the electric dipole moment operator we have $\langle \bar{r}|\, \boldsymbol{\mu}\,|\bar{s}\rangle = \langle r|\, \boldsymbol{\mu}\,|s\rangle$, since the unitary operator U commutes with $\boldsymbol{\mu}$. Similarly $\langle \bar{r}|\, \bar{\boldsymbol{m}}\,|\bar{s}\rangle = \langle r\,|\, \boldsymbol{m}\,|s\rangle$, where the magnetic dipole moment operators (which should not be confused with the magnetization field $\boldsymbol{m}$ and $\tilde{\boldsymbol{m}}$) are defined by $\bar{\boldsymbol{m}} = -(e/mc)\sum_\alpha \bar{\boldsymbol{l}}_\alpha$ and $\boldsymbol{m} = (-e/mc)\sum_\alpha \boldsymbol{l}_\alpha$ and the angular momentum operators by $\bar{\boldsymbol{l}}_\alpha = (\boldsymbol{q}_\alpha - \boldsymbol{R}) \times \bar{\boldsymbol{p}}_\alpha$ and $\boldsymbol{l}_\alpha = (\boldsymbol{q}_\alpha - \boldsymbol{R}) \times \boldsymbol{p}_\alpha$, so that $\bar{\boldsymbol{l}}_\alpha = U\boldsymbol{l}_\alpha U^{-1}$ and $\bar{\boldsymbol{m}} = U\boldsymbol{m}U^{-1}$.

The use of the minimal-coupling and multipolar Hamiltonians is calculating spectral line shapes when the "energy shell" approximations leading to the Lorentzian profile of Section 6.3 are not valid has been discussed by Power and Zienau (1959) and, more recently, by Davidovich and Nussenzveig (1980).

REFERENCES

Atkins, P. W. and Woolley, R. G. (1970). "The interaction of molecular multipoles with the electromagnetic field in the canonical formulation of non-covariant quantum electrodynamics". *Proc. Roy. Soc. Lond. A* **319,** 549.

Babiker, M. (1975). "Bound state quantum electrodynamics with polarization sources". *Proc. Roy. Soc. Lond. A* **342,** 113.

Babiker, M., Power, E. A. and Thirunamachandran, T. (1973). "Atomic field equations for Maxwell fields interacting with non-relativistic quantal sources". *Proc. Roy. Soc. Lond. A* **332,** 187.

Babiker, M., Power, E. A. and Thirunamachandran, T. (1974). "On a generalization of the Power–Zienau–Woolley transformation in quantum electrodynamics and atomic field equations". *Proc. Roy. Soc. Lond. A* **338,** 235.

Davidovich, L. and Nussenzveig, H. M. (1980). *In* "Foundations of Radiation Theory and Quantum Electrodynamics" (ed. A. O. Barut). Plenum Press, New York and London.

de Groot, S. R. (1969). "The Maxwell Equations". North Holland, Amsterdam.

Felderhof, B. U. and Adu-Gyamfi, D. (1974). "The multipole Hamiltonian for molecules". *Physica* **71,** 399.

Fiutak, T. (1963). "The multipole expansion in quantum theory". *Can. J. Phys.* **41,** 12.

Göppert-Mayer, M. (1931). "Über Elementarakte mit zwei Quantensprüngen". *Ann. Phys. Leipzig* **9,** 273.

Healy, W. P. (1977a). "Centre of mass motion in non-relativistic quantum electrodynamics". *J. Phys. A: Math. Gen.* **10,** 279.

Healy, W. P. (1977b). "The representation of microscopic charge and current densities in terms of polarization and magnetization fields". *Proc. Roy. Soc. Lond. A* **358,** 367.

Healy, W. P. (1977c). "A generalization of the Kramers–Heisenberg dispersion formula". *Phys. Rev. A* **16,** 1568.

Healy, W. P. (1978). "Covariant representation of microscopic charge and current densities in terms of polarization and magnetization fields". *J. Phys. A: Math. Gen.* **11,** 1899.

Healy, W. P. (1979). "Line integral Lagrangians for the electromagnetic fields interacting with charged particles". *Phys. Rev. A* **19,** 2353.

Healy, W. P. (1980). "Path-dependent Lagrangians in relativistic electrodynamics". *J. Phys. A: Math. Gen.* **13,** 2383.

Healy, W. P. and Woolley, R. G. (1978). "On the derivation of the Kramers–Heisenberg dispersion formula from non-relativistic quantum electrodynamics". *J. Phys. B: Atom. Molec. Phys.* **11,** 1131.

Lanczos, C. (1970). "The Variational Principles of Mechanics". University of Toronto Press.

Power, E. A. (1964). "Introductory Quantum Electrodynamics". Longmans, London.
Power, E. A. (1978). *In* "Multiphoton Processes" (ed. J. H. Eberly and P. Lambropoulos). Wiley, New York.
Power, E. A. and Thirunamachandran, T. (1971). "Three distribution identities and Maxwell's atomic field equations including the Röntgen current". *Mathematika* **18,** 240.
Power, E. A. and Thirunamachandran, T. (1978). "On the nature of the Hamiltonian for the interaction of radiation with atoms and molecules: $(e/mc)\mathbf{p}\,.\,\mathbf{A}, -\boldsymbol{\mu}\,.\,\boldsymbol{E}$, and all that". *Am. J. Phys.* **46,** 370.
Power, E. A. and Zienau, S. (1959). "Coulomb gauge in non-relativistic quantum electrodynamics and the shape of spectral lines". *Phil. Trans. Roy. Soc. Lond. A* **251,** 427.
Richards, P. I. (1948). "On the Hamiltonian for a particle in an electromagnetic field". *Phys. Rev.* **73,** 254.
Whittaker, E. T. (1961). "A Treatise on the Analytical Dynamics of Particles and Rigid Bodies". Cambridge University Press.
Woolley, R. G. (1971a). "Molecular quantum electrodynamics". *Proc. Roy. Soc. Lond. A* **321,** 557.
Woolley, R. G. (1971b). "On the Hamiltonian theory of the molecule–electromagnetic field system". *Molec. Phys.* **22,** 1013.
Woolley, R. G. (1975a). "The electrodynamics of atoms and molecules". *Adv. Chem. Phys.* **33,** 153.
Woolley, R. G. (1975b). "On non-relativistic electron theory". *Ann. Inst. Henri Poincaré* **23,** 365.
Woolley, R. G. and Cordle, J. E. (1973). "On the gauge invariance of the Schrödinger equation". *Chem. Phys. Lett.* **22,** 411.

Appendix A

Values of Physical Constants

Planck's constant	$h = 6{\cdot}626 \times 10^{-27}$ erg sec	$6{\cdot}626 \times 10^{-34}$ J sec
	$\hbar = 1{\cdot}055 \times 10^{-27}$ erg sec	$1{\cdot}055 \times 10^{-34}$ J sec
Speed of light *in vacuo*	$c = 2{\cdot}998 \times 10^{10}$ cm sec^{-1}	$2{\cdot}998 \times 10^{8}$ m sec^{-1}
Boltzmann's constant	$k = 1{\cdot}381 \times 10^{-16}$ erg K^{-1}	$1{\cdot}381 \times 10^{-23}$ J K^{-1}
Charge of electron	$-e = -4{\cdot}803 \times 10^{-10}$ esu	$1{\cdot}603 \times 10^{-19}$ C
Mass of electron	$m = 9{\cdot}110 \times 10^{-28}$ g	$9{\cdot}110 \times 10^{-31}$ kg
Mass of proton	$M = 1{\cdot}673 \times 10^{-24}$ g	$1{\cdot}673 \times 10^{-27}$ kg
Fine structure constant	$\alpha = e^2/(\hbar c) = 7{\cdot}297 \times 10^{-3} \approx 1/137$	—
Bohr radius	$a_0 = \hbar^2/(me^2) = 5{\cdot}292 \times 10^{-9}$ cm	$5{\cdot}292 \times 10^{-11}$ m
Angstrom unit	$1\ \text{Å} = 10^{-8}$ cm	10^{-10} m

Appendix B

Theorems in Vector Analysis

In this appendix we state several theorems in vector analysis that have been used extensively in the body of the book, particularly in Chapter 7. It has been assumed throughout that the normal to a closed surface always points outwards, whether the coordinate system is right- or left-handed, and that any volume element is a true scalar and positive. Similarly, every curve is assigned a sense that is independent of the chirality of the reference frame. The sense of an open surface, however, is determined from that of its closed boundary curve by the right-hand rule in a right-handed frame and the left-hand rule in a left-handed frame, so that the directed surface is a pseudovector. The algebraic and differential operators on vectors have their usual meanings; thus $\nabla \times \boldsymbol{F}$ is a pseudovector, if $\boldsymbol{F}$ is a true vector. With these conventions the following well known theorems are valid in either type of frame.

THEOREM 1 (Stokes' theorem)

$$\oint_C \boldsymbol{F} \cdot \mathrm{d}\boldsymbol{x} = \iint_S \nabla \times \boldsymbol{F} \cdot \mathrm{d}\boldsymbol{S} \tag{B.1}$$

THEOREM 2 (the divergence theorem)

$$\oiint_S \boldsymbol{F} \cdot \mathrm{d}\boldsymbol{S} = \iiint_V \nabla \cdot \boldsymbol{F}\, \mathrm{d}^3 x \tag{B.2}$$

THEOREM 3

$$\oint_C f\, \mathrm{d}\boldsymbol{x} = \iint_S \mathrm{d}\boldsymbol{S} \times \nabla f \tag{B.3}$$

THEOREM 4

$$\oint_C \mathrm{d}\boldsymbol{x} \times \boldsymbol{F} = \iint_S (\mathrm{d}\boldsymbol{S} \times \nabla) \times \boldsymbol{F} \tag{B.4}$$

THEOREM 5

$$\oiint_S f\,\mathrm{d}\boldsymbol{S} = \iiint_V \nabla f\,\mathrm{d}^3x \tag{B.5}$$

THEOREM 6

$$\oiint_S \mathrm{d}\boldsymbol{S}\times\boldsymbol{F} = \iiint_V \nabla\times\boldsymbol{F}\,\mathrm{d}^3x \tag{B.6}$$

Here f and $\boldsymbol{F}$ are differentiable scalar and vector fields, respectively, and may depend on the time t as well as on the point $\boldsymbol{x}$. In Theorems 1, 3, and 4, C is a closed curve bounding an open surface S. In Theorems 2, 5 and 6, S is a closed surface bounding a volume V. Theorems 3 and 4 may easily be derived from Stokes' theorem and Theorems 5 and 6 from the divergence theorem.

The remaining theorems (Healy 1977) have to do with moving curves, surfaces or volumes and the proofs of these will be given. We consider first two moving points that have definite and smooth trajectories $\boldsymbol{x}_1(t)$ and $\boldsymbol{x}_2(t)$ and suppose them joined at any instant t by a curve C_{12} that is traced out in the positive sense by going along it from $\boldsymbol{x}_1$ to $\boldsymbol{x}_2$. To describe the motion of the intermediate points, the curve will be taken to be specified by a real parameter u varying in the domain u_1 to u_2, with u_1 corresponding to $\boldsymbol{x}_1$ and u_2 to $\boldsymbol{x}_2$. Notwithstanding that C_{12} may change its length as time proceeds, the parameter u can always be scaled so that u_1 and u_2 are independent of time, and this will be supposed done. The points $\boldsymbol{x}'$ of C_{12} are thus determined by a differentiable function $\boldsymbol{x}'(u, t)$ of the two independent variables u and t. Each point of C_{12} is associated for all time with a fixed value of u and will be assumed to describe a smooth path as t varies, the velocity at any point of such a path being $\dot{\boldsymbol{x}}'$. C_{12} itself then sweeps out a smooth surface bounded by the trajectories of $\boldsymbol{x}_1$ and $\boldsymbol{x}_2$. With these preliminaries disposed of we can now state

THEOREM 7

$$\dot{\boldsymbol{x}}_2\,\delta(\boldsymbol{x}-\boldsymbol{x}_2)-\dot{\boldsymbol{x}}_1\,\delta(\boldsymbol{x}-\boldsymbol{x}_1)=\frac{\partial}{\partial t}\int_{C_{12}}\mathrm{d}\boldsymbol{x}'\,\delta(\boldsymbol{x}-\boldsymbol{x}')+\nabla\times\int_{C_{12}}\mathrm{d}\boldsymbol{x}'\times\dot{\boldsymbol{x}}'\,\delta(\boldsymbol{x}-\boldsymbol{x}') \tag{B.7}$$

To prove it we note that

$$\begin{aligned}\frac{\partial}{\partial t}\int_{C_{12}}\mathrm{d}x_i'\,\delta(\boldsymbol{x}-\boldsymbol{x}') &= \frac{\partial}{\partial t}\int_{u_1}^{u_2}\mathrm{d}u\,\frac{\partial x_i'}{\partial u}\,\delta(\boldsymbol{x}-\boldsymbol{x}')\\ &= \int_{u_1}^{u_2}\mathrm{d}u\left(\frac{\partial \dot{x}_i'}{\partial u}+\frac{\partial x_i'}{\partial u}\frac{\partial}{\partial t}\right)\delta(\boldsymbol{x}-\boldsymbol{x}')\end{aligned} \tag{B.8}$$

and that

$$\nabla \times \int_{C_{12}} d\boldsymbol{x}' \times \dot{\boldsymbol{x}}' \, \delta(\boldsymbol{x}-\boldsymbol{x}') \, |_i = \int_{u_1}^{u_2} du \left(\dot{x}_i' \frac{\partial x_j'}{\partial u} - \frac{\partial x_i'}{\partial u} \dot{x}_j' \right) \nabla_j' \, \delta(\boldsymbol{x}-\boldsymbol{x}')$$

$$= \int_{u_1}^{u_2} du \left(\dot{x}_i' \frac{\partial}{\partial u} - \frac{\partial x_i'}{\partial u} \frac{\partial}{\partial t} \right) \delta(\boldsymbol{x}-\boldsymbol{x}')$$

$$= \dot{x}_{2i} \, \delta(\boldsymbol{x}-\boldsymbol{x}_2) - \dot{x}_{1i} \, \delta(\boldsymbol{x}-\boldsymbol{x}_1)$$

$$- \int_{u_1}^{u_2} du \left(\frac{\partial \dot{x}_i'}{\partial u} + \frac{\partial x_i'}{\partial u} \frac{\partial}{\partial t} \right) \delta(\boldsymbol{x}-\boldsymbol{x}') \qquad \text{(B.9)}$$

where the last step involves an integration by parts. Addition of Equations (B.8) and (B.9) gives the ith component of Equation (B.7).

We consider next an open surface S moving through space in a continuous manner. Its area and shape may vary with time, but the surface is nevertheless to be specified by two real parameters u and v varying in the domains u_1 to u_2 and v_1 to v_2, with these limits being time-independent. The points $\boldsymbol{x}'$ of S are thus determined by a differentiable function $\boldsymbol{x}'(u, v, t)$ of the three independent variables u, v and t. We now prove

THEOREM 8

$$\frac{\partial}{\partial t} \iint_S d\boldsymbol{S}' \, \delta(\boldsymbol{x}-\boldsymbol{x}') + \nabla \iint_S d\boldsymbol{S}' \cdot \dot{\boldsymbol{x}}' \, \delta(\boldsymbol{x}-\boldsymbol{x}') = -\iint_S (d\boldsymbol{S}' \times \nabla') \times \{\dot{\boldsymbol{x}}' \, \delta(\boldsymbol{x}-\boldsymbol{x}')\} \qquad \text{(B.10)}$$

We have first that

$$d\boldsymbol{S}' = du \, dv \frac{\partial \boldsymbol{x}'}{\partial u} \times \frac{\partial \boldsymbol{x}'}{\partial v} \qquad \text{(B.11)}$$

the labels of the variables u and v being interchanged if necessary to give $d\boldsymbol{S}'$ its proper sense. We have also

$$d\boldsymbol{S}' \times \nabla' = du \, dv \left\{ \frac{\partial \boldsymbol{x}'}{\partial v} \left(\frac{\partial \boldsymbol{x}'}{\partial u} \cdot \nabla' \right) - \frac{\partial \boldsymbol{x}'}{\partial u} \left(\frac{\partial \boldsymbol{x}'}{\partial v} \cdot \nabla' \right) \right\}$$

$$= du \, dv \left(\frac{\partial \boldsymbol{x}'}{\partial v} \frac{\partial}{\partial u} - \frac{\partial \boldsymbol{x}'}{\partial u} \frac{\partial}{\partial v} \right) \qquad \text{(B.12)}$$

This shows that $d\boldsymbol{S}' \times \nabla'$ operates along the surface S, which must be so if the integral on the right-hand side of Equation (B.10) is to be meaningful, since at any instant $\dot{\boldsymbol{x}}'$ is defined only on S. Using Equations (B.11) and (B.12) and the antisymmetric property of the Levi–Civita tensor, we obtain after some rearrangement and cancellation the equation

$$\left[\frac{\partial}{\partial t}\iint\limits_{S} \mathrm{d}\boldsymbol{S}'\,\delta(\boldsymbol{x}-\boldsymbol{x}')+\iint\limits_{S}(\mathrm{d}\boldsymbol{S}'\times\nabla')\times\{\dot{\boldsymbol{x}}'\,\delta(\boldsymbol{x}-\boldsymbol{x}')\}\right]_i$$

$$=\varepsilon_{ijk}\int_{u_1}^{u_2}\int_{v_1}^{v_2}\mathrm{d}u\,\mathrm{d}v\left\{\frac{\partial x_j'}{\partial u}\frac{\partial x_k'}{\partial v}\frac{\partial}{\partial t}-\dot{x}_j'\left(\frac{\partial x_k'}{\partial v}\frac{\partial}{\partial u}-\frac{\partial x_k'}{\partial u}\frac{\partial}{\partial v}\right)\right\}\delta(\boldsymbol{x}-\boldsymbol{x}')$$

$$=\iint\limits_{S}\{\mathrm{d}S_i(\dot{\boldsymbol{x}}'\,.\,\nabla')-[\dot{\boldsymbol{x}}'\times(\mathrm{d}\boldsymbol{S}'\times\nabla')]_i\}\,\delta(\boldsymbol{x}-\boldsymbol{x}')$$

$$=-\nabla_i\iint\limits_{S}\mathrm{d}\boldsymbol{S}'\,.\,\dot{\boldsymbol{x}}'\,\delta(\boldsymbol{x}-\boldsymbol{x}') \tag{B.13}$$

which is equivalent to the ith component of Equation (B.10).

We consider finally a moving volume V bounded by a closed surface and the points of which are specified at any time by three real parameters u, v and w varying in the domains u_1 to u_2, v_1 to v_2 and w_1 to w_2. These limits are again taken to be time-independent, although V may change in volume and shape. The velocity of an interior point of V is $\dot{\boldsymbol{x}}'$, where $\boldsymbol{x}'(u, v, w, t)$ is a differentiable function of the four independent variables u, v, w and t. For this moving volume we have the following theorem.

THEOREM 9

$$\iiint\limits_{V}\mathrm{d}^3x'\nabla'\,.\,\{\dot{\boldsymbol{x}}'\delta(\boldsymbol{x}-\boldsymbol{x}')\}=\frac{\partial}{\partial t}\iiint\limits_{V}\mathrm{d}^3x'\,\delta(\boldsymbol{x}-\boldsymbol{x}') \tag{B.14}$$

Now the differential volume element appearing in Equation (B.14) is given by

$$\mathrm{d}^3x'=\pm\mathrm{d}u\,\mathrm{d}v\,\mathrm{d}w\frac{\partial\boldsymbol{x}'}{\partial u}\times\frac{\partial\boldsymbol{x}'}{\partial v}\,.\,\frac{\partial\boldsymbol{x}'}{\partial w} \tag{B.15}$$

with the labels u, v and w being permuted if necessary to make d^3x' positive. The minus sign in Equation (B.15) applies in a left-handed frame if the plus sign applies in a right-handed frame. (Note that u, v and

w are assumed to be invariant parameters.) We have also

$$\frac{\partial}{\partial t}\left(\frac{\partial \boldsymbol{x}'}{\partial u}\times\frac{\partial \boldsymbol{x}'}{\partial v}\cdot\frac{\partial \boldsymbol{x}'}{\partial w}\right)=\varepsilon_{ljk}\frac{\partial \dot{x}'_l}{\partial u}\frac{\partial x'_j}{\partial v}\frac{\partial x'_k}{\partial w}+\varepsilon_{imk}\frac{\partial x'_i}{\partial u}\frac{\partial \dot{x}'_m}{\partial v}\frac{\partial x'_k}{\partial w}+\varepsilon_{ijn}\frac{\partial x'_i}{\partial u}\frac{\partial x'_j}{\partial v}\frac{\partial \dot{x}'_n}{\partial w}$$

$$=\frac{\partial x'_i}{\partial u}\frac{\partial x'_j}{\partial v}\frac{\partial x'_k}{\partial w}(\varepsilon_{ljk}\nabla'_i\dot{x}'_l+\varepsilon_{imk}\nabla'_j\dot{x}'_m+\varepsilon_{ijn}\nabla'_k\dot{x}'_n) \qquad \text{(B.16)}$$

The expression in brackets on the right-hand side changes sign if any two of i, j and k are interchanged and is therefore proportional to ε_{ijk}. Contracting this expression with ε_{ijk} and using the properties (B.17),

$$\varepsilon_{ijk}\varepsilon_{ijl}=2\delta_{kl} \qquad \varepsilon_{ijk}\varepsilon_{ijk}=6 \qquad \text{(B.17)}$$

we see that the factor of proportionality is $\nabla'\cdot\dot{\boldsymbol{x}}'$. Thus

$$\frac{\partial}{\partial t}\iiint_V \mathrm{d}^3x'\,\delta(\boldsymbol{x}-\boldsymbol{x}')$$

$$=\pm\int_{u_1}^{u_2}\int_{v_1}^{v_2}\int_{w_1}^{w_2}\mathrm{d}u\,\mathrm{d}v\,\mathrm{d}w\left\{\frac{\partial}{\partial t}\left(\frac{\partial \boldsymbol{x}'}{\partial u}\times\frac{\partial \boldsymbol{x}'}{\partial v}\cdot\frac{\partial \boldsymbol{x}'}{\partial w}\right)+\left(\frac{\partial \boldsymbol{x}'}{\partial u}\times\frac{\partial \boldsymbol{x}'}{\partial v}\cdot\frac{\partial \boldsymbol{x}'}{\partial w}\right)\frac{\partial}{\partial t}\right\}\delta(\boldsymbol{x}-\boldsymbol{x}')$$

$$=\pm\int_{u_1}^{u_2}\int_{v_1}^{v_2}\int_{w_1}^{w_2}\mathrm{d}u\,\mathrm{d}v\,\mathrm{d}w\left(\frac{\partial \boldsymbol{x}'}{\partial u}\times\frac{\partial \boldsymbol{x}'}{\partial v}\cdot\frac{\partial \boldsymbol{x}'}{\partial w}\right)(\nabla'\cdot\dot{\boldsymbol{x}}'+\dot{\boldsymbol{x}}'\cdot\nabla')\,\delta(\boldsymbol{x}-\boldsymbol{x}')$$

$$=\iiint_V \mathrm{d}^3x'\nabla'\cdot\{\dot{\boldsymbol{x}}'\,\delta(\boldsymbol{x}-\boldsymbol{x}')\} \qquad \text{(B.18)}$$

which proves the theorem.

Appendix C

Longitudinal and Transverse Vector Fields

The division of a vector field $\boldsymbol{v}(\boldsymbol{x})$ into irrotational and solenoidal parts corresponds to the resolution of its Fourier transform $\tilde{\boldsymbol{v}}(\boldsymbol{k})$ into components parallel and perpendicular to $\boldsymbol{k}$ (Belinfante 1946, Power 1964). If $\hat{\boldsymbol{k}}$ is the unit vector in the direction of $\boldsymbol{k}$, the identity

$$\tilde{\boldsymbol{v}} = (\hat{\boldsymbol{k}} \cdot \tilde{\boldsymbol{v}})\hat{\boldsymbol{k}} + (\hat{\boldsymbol{k}} \times \tilde{\boldsymbol{v}}) \times \hat{\boldsymbol{k}} \tag{C.1}$$

implies that

$$\boldsymbol{v}(\boldsymbol{x}) = \boldsymbol{v}^{\parallel}(\boldsymbol{x}) + \boldsymbol{v}^{\perp}(\boldsymbol{x}) \tag{C.2}$$

where

$$\boldsymbol{v}^{\parallel}(\boldsymbol{x}) = \frac{1}{(2\pi)^{3/2}} \iiint (\hat{\boldsymbol{k}} \cdot \tilde{\boldsymbol{v}})\hat{\boldsymbol{k}} e^{i\boldsymbol{k}\cdot\boldsymbol{x}}\, \mathrm{d}^3k \tag{C.3}$$

and is an irrotational field, and where

$$\boldsymbol{v}^{\perp}(\boldsymbol{x}) = \frac{1}{(2\pi)^{3/2}} \iiint (\hat{\boldsymbol{k}} \times \tilde{\boldsymbol{v}}) \times \hat{\boldsymbol{k}} e^{i\boldsymbol{k}\cdot\boldsymbol{x}}\, \mathrm{d}^3k \tag{C.4}$$

and is a solenoidal field. Because of this, $\boldsymbol{v}^{\parallel}$ is known as the longitudinal part and $\boldsymbol{v}^{\perp}$ as the transverse part of $\boldsymbol{v}$. The decomposition (C.2) can also be effected by using longitudinal and transverse delta-function dyadics defined by

$$\begin{aligned}\delta^{\parallel}_{ij}(\boldsymbol{x}) &= \frac{1}{(2\pi)^3} \iiint \hat{k}_i \hat{k}_j e^{i\boldsymbol{k}\cdot\boldsymbol{x}}\, \mathrm{d}^3k \\ &= -\frac{1}{4\pi} \nabla_i \nabla_j \frac{1}{x}\end{aligned} \tag{C.5}$$

and

$$\begin{aligned}\delta^{\perp}_{ij}(\boldsymbol{x}) &= \frac{1}{(2\pi)^3} \iiint (\delta_{ij} - \hat{k}_i \hat{k}_j) e^{i\boldsymbol{k}\cdot\boldsymbol{x}}\, \mathrm{d}^3k \\ &= \delta_{ij}\, \delta(\boldsymbol{x}) + \frac{1}{4\pi} \nabla_i \nabla_j \frac{1}{x}\end{aligned} \tag{C.6}$$

Here δ_{ij} is the Kronecker delta. Equations (C.3) and (C.4) are then equivalent to

$$v_i^{\parallel}(\boldsymbol{x}) = \iiint \delta_{ij}^{\parallel}(\boldsymbol{x}-\boldsymbol{x}')v_j(\boldsymbol{x}')\,\mathrm{d}^3x' \tag{C.7}$$

and

$$v_i^{\perp}(\boldsymbol{x}) = \iiint \delta_{ij}^{\perp}(\boldsymbol{x}-\boldsymbol{x}')v_j(\boldsymbol{x}')\,\mathrm{d}^3x' \tag{C.8}$$

We have also

$$\delta_{ij}\,\delta(\boldsymbol{x}) = \delta_{ij}^{\parallel}(\boldsymbol{x}) + \delta_{ij}^{\perp}(\boldsymbol{x}) \tag{C.9}$$

from which Equation (C.2) may be recovered.

The irrotational and solenoidal parts of $\boldsymbol{v}$ are not uniquely defined, since the gradient of a harmonic function could be added to $\boldsymbol{v}^{\parallel}$ and subtracted from $\boldsymbol{v}^{\perp}$. The partitioning can be made unique, however, by requiring that $\boldsymbol{v}^{\parallel}$ and $\boldsymbol{v}^{\perp}$, as well as $\boldsymbol{v}$, tend to zero sufficiently rapidly at infinity. We have in any case assumed that the Fourier transforms of both $\boldsymbol{v}^{\perp}$ and $\boldsymbol{v}^{\parallel}$ and of their first derivatives exist, and that the Fourier inversion theorem holds.

Appendix D

Operator Identities

Two useful identities satisfied by linear operators A and B are derived in the following theorems (Messiah 1962). It is assumed that A and B act on the same vector space and that all products and series that occur are well defined.

THEOREM 1

$$e^{A}Be^{-A} = B + [A, B] + \frac{1}{2!}[A, [A, B]] + \dots . \tag{D.1}$$

Proof. Let $C(\tau) = e^{A\tau}Be^{-A\tau}$, where τ is a real variable. Then

$$\frac{\mathrm{d}C}{\mathrm{d}\tau} = [A, C(\tau)] \tag{D.2}$$

or, since $C(0) = B$,

$$C(\tau) = B + \int_0^{\tau} [A, C(\tau_1)]\,\mathrm{d}\tau_1 \tag{D.3}$$

By successive iterations we obtain

$$\begin{aligned} C(\tau) &= B + [A, B]\int_0^{\tau} \mathrm{d}\tau_1 + [A, [A, B]]\int_0^{\tau} \mathrm{d}\tau_1 \int_0^{\tau_1} \mathrm{d}\tau_2 + \dots \\ &= B + [A, B]\tau + \frac{1}{2!}[A, [A, B]]\tau^2 + \dots . \end{aligned} \tag{D.4}$$

The integral in the general term represents the volume of the $1/n!$ part of an n-dimensional rectangular parallelepiped of side τ. Thus

$$\int_0^{\tau} \mathrm{d}\tau_1 \int_0^{\tau_1} \mathrm{d}\tau_2 \dots \int_0^{\tau_{n-1}} \mathrm{d}\tau_n = \frac{1}{n!}\tau^n \qquad n = 1, 2, \dots \tag{D.5}$$

as may be shown by induction. Putting $\tau = 1$ in Equation (D.4) gives the required result.

THEOREM 2. If $[A, B]$ commutes with both A and B, then

$$e^A e^B = e^{A+B} e^{1/2[A,B]} \tag{D.6}$$

This holds, in particular, when $[A, B]$ is a c-number.

Proof. Let $C(\tau) = e^{A\tau} e^{B\tau}$. Then

$$\begin{aligned} \frac{\mathrm{d}C}{\mathrm{d}\tau} &= (A + e^{A\tau} B e^{-A\tau}) C(\tau) \\ &= (A + B + [A, B]\tau) C(\tau) \end{aligned} \tag{D.7}$$

where the second line follows from Theorem 1 and the hypothesis that $[A, [A, B]] = 0$. Since also $[B, [A, B]] = 0$ and $C(0) = 1$, we have

$$C(\tau) = e^{(A+B)\tau} e^{1/2[A,B]\tau^2} \tag{D.8}$$

as may be verified by differentiation. Putting $\tau = 1$ then gives the required result.

REFERENCES FOR APPENDICES A–D

Belinfante, F. J. (1946). "On the longitudinal and the transversal delta-function, with some applications". *Physica* **12,** 1.

Healy, W. P. (1977). "The representation of microscopic charge and current densities in terms of polarization and magnetization fields". *Proc. Roy. Soc. Lond.* **A358,** 367.

Messiah, A. (1962). "Quantum Mechanics". North Holland, Amsterdam.

Power, E. A. (1964). "Introductory Quantum Electrodynamics". Longmans, London.

Subject Index